Ideas
and
Information

Ideas and Information

Managing in a High-Tech World

Arno Penzias

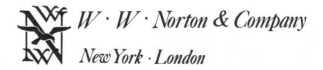
W · W · Norton & Company

New York · London

FIRST EDITION

THE TEXT OF THIS BOOK *is composed in Times Roman, with display type set in Windsor
and Caslon. Composition and manufacturing by The Haddon Craftsmen, Inc. Book
design by Marjorie J. Flock.*

Library of Congress Cataloging-in-Publication Data

Penzias, Arno A.
    Ideas and information : managing in a high-tech world / Arno
Penzias.
    p.   cm.
    Includes index.
    1. Computers and civilization.   I. Title.
QA76.9.C66P46 1989
303.4'834—dc19   88-14180

ISBN 0-393-02649-3

W. W. Norton & Company, Inc., 500 Fifth Avenue, New York, N. Y. 10110
W. W. Norton & Company Ltd., 37 Great Russell Street, London WC1B 3NU

1 2 3 4 5 6 7 8 9 0

To Anne Barras Penzias,
David, Mindy, Laurie, Bonnie, and Rachel,
with love

Contents

Preface

IN MY JOB at AT&T Bell Laboratories, I have spoken to a great many groups over the past few years—graduating high school seniors, human resources managers, boards of directors of major institutions—groups of all kinds, all over the United States and in other parts of the world as well. Despite their different backgrounds and points of view, several common threads of interest unite them.

First, they share a curiosity about the computer technology that surrounds us, how it works, and its potential for "intelligence."

Second, they are generally more than a little uneasy about the impact of technology on their lives and on the world they live in.

Finally, they want to know about the future. How will they manage in the world of technology to come?

I hope that this book will answer some of these questions and concerns, thereby making its readers more comfortable with technology and better able to profit from it.

Getting the best from the combination of mind and machinery requires an understanding of how they fit together and the roles that each might play. I have tried to write for those who don't bring technical knowledge to their reading, so I give less attention to engineering and mathematical applications of computing, for example, than to capabilities that a nontechnical user might encounter someday (such as computer understanding of everyday language).

There's more to information technology than the computer itself. In addition to using modern tools—such as telephones, microfilm, and computers—to help us transmit, store, and process information, we humans employ a great deal of older technology as well. Thanks to the inventions of writing, drawing, and arithmetic, for example, a "simple" piece of paper can be used to send a message, preserve an image, or aid a calculation. Together, all these technologies supplement the human mind's ability to communicate, remember, and think.

I want to demystify information processing by giving you a look at what goes on beneath the slick surface layer that computers often present to their potential users. Beneath that layer, computers have much in common with their more easily understandable ancestors—notched sticks, clay tablets, clocks, and adding machines.

We can best understand information processing as the manipulation of symbols—particles of thought, like numbers, words, and pictures—that are subject to the rules of logic, grammar, and arithmetic which both humans and computers employ. These rules enable computers to deal with problems that humans solve by using intelligence, such as recognizing handwriting, winning a chess game, or diagnosing problems in malfunctioning machinery. Still, even the best of present-day computing systems lack a key attribute of intelligence: the ability to move from one context to another.

Even though a computer's actions sometimes mimic human intelligence, such machines are fundamentally different from brains. While computers afford humans much valuable help in processing massive amounts of information—as in refining the shape of a jetliner's wings, or keeping the books of the Social

Security Administration—they offer little serious competition in the areas of creativity, integration of disparate information, and flexible adaptation to unforeseen circumstances. Here the human mind functions best.

Unlike machines, human minds can create ideas. We need ideas to guide us toward progress, as well as tools to implement them. As with any other set of tools, the real power of information technology comes from the human ideas that create and focus it.

But as technology helps us shape our environment, it also affects the way we live and work. With the steady increase in human knowledge, our employment of technology continues to enhance human specialization and increases the interdependency between communities of specialized knowledge. I speculate that many large organizations may therefore grow even larger in order to bring diverse knowledge to bear on complex tasks. But this trend needn't dehumanize our lives. As I see it, properly directed technology can enhance the role of the individual by giving each of us needed access to heretofore inaccessible information sources. In this way, frustrating encounters with isolated pockets of incomplete and misleading information may be replaced by one-to-one interactions between human beings empowered by supportive technology.

In all this, I have no doubt that the world's most powerful information tool will continue to be the human mind.

Acknowledgments

FIRST AND FOREMOST, the book owes its existence
to Dan Stern's persistent persuasion. Dan decided that
I had something to write about, would find the time
to write it, and would enjoy working with Ed Barber—
almost a year before I thought enough of the idea to
scribble my first outline. In addition to Dan's efforts,
I benefited from helpful advice and early encouragement from Bill Leigh, Laurens Schwartz, and Lillian
Schwartz. I'm also grateful to Al Aho, Bill Bucci, Jon
Bentley, Ron Brachman, Sandy Fraser, Paul Henry,
Donna Kelly, Marilyn Kappel, Marilyn Laurie, Bob
Martin, and David Penzias for help in improving the
manuscript with their comments and corrections.

Most of the material contained in the book itself reflects the time and effort generously provided by my Bell
Labs colleagues in helping to educate me over the past
twenty-seven years. While some of their names are listed
in the text itself, I am equally grateful to many others
for the day-to-day interactions that make working at
Bell Labs such a rewarding experience.

Finally, I owe special debts of gratitude to Ed Barber
and Julia Heiney. With patience, dedication, unfailing
good humor, and eloquent examples, Ed waged a year-
long battle against prolixity, vague generality, and a
whole list of bad writing habits I didn't know I had. As

each new draft came back doused in red ink, Julia would once again convert my penciled revisions into neat rows of word-processed characters, all the while orchestrating my work life and the demands of a busy office around my writing excursions.

Many thanks to all.

Ideas
and
Information

Chapter 1

Information Processing

Where is the information we have lost in data?
— HIROSHI INOSE and J. R. PIERCE, *Information Technology and Civilization*

I ALMOST missed getting my Ph.D. because I couldn't extract the information I needed from the data I had gathered. As a graduate student, I had chosen to work in radio astronomy, to "listen" for clues to the physical workings of distant galaxies. Thus, I spent the fall of 1960 probing the radio spectrum of a cluster of galaxies located in the constellation Pegasus—vainly, it seemed, attempting to locate traces of intergalactic hydrogen gas. Hydrogen is the simplest and most abundant element in the universe and the fundamental material from which galaxies are formed. Other radio astronomers had detected hydrogen *within* some galaxies, but now I was searching for the gas that theory predicted would lie in the space *between* them. My thesis project hung on the outcome of that search.

I had mounted my hand-crafted receiver at the focus of a radio antenna borrowed from the U.S. Naval Research Laboratory at a remote site in southern Maryland. That receiver represented three years of painstaking labor invested in the hope that the few weeks of observing time allotted to me would produce a meaningful result.

With everything—electronics, cryogenic cooling system, and microwave components—working properly, I would tune my receiver to the proper frequency, point the antenna to a spot in the sky that my galaxies would reach in half an hour, lock the antenna at that place, and wait for the rotation of the earth to bring the galaxies into target position.* A clock motor drove a roll of chart paper past a pen connected to the output of my receiver, thereby producing a five-foot-long wiggly line for each observation of an hour or so. A signal from intergalactic hydrogen (or any other source of radio energy at the same wavelength) detected by the receiver would deflect the pen and thereby create a bump in that line at the time the galaxies passed through the antenna's field of view. No such bumps showed up on any of my scans. The only evident wiggles in each record were due to random noise fluctuations in the receiver.

As the days passed, my pile of charts rose ever higher. But I was, in fact, working almost blind because I could look at only one scan at a time—never combine the data from one with those of another—and none of the individual scans revealed the presence of a signal. Giving up on the hope of discovering a massive amount of hydrogen, I changed the ground rules. I doggedly repeated the same observation over and over, hoping to lower the threshold of sensitivity by averaging a large number of scans together and thereby setting a truly sensitive upper limit to the possible presence of the gas. (Averaging enhances the visibility of weak signals, because the random fluctuations due to noise tend to cancel out from one scan to another.) I pressed on with my observations until the last possible minute, because I knew

*Moving the antenna during an observation would have disturbed the receiver too much. Instead, my observing method depended on the earth's rotation to sweep the receiving antenna's beam across the region of interest.

I wouldn't get another chance. Much of my apparatus had been soldered together at the site. Once dismantled, reassembly would have meant almost starting from scratch.

When my observing session ended, I spent the next weeks reducing my data—carefully measuring the pen positions of these dozens of long documents, at six-second intervals along each hour-long scan, and then averaging all the scans together. To my surprise, the final results showed a small but definite bump at just the spot where the intergalactic hydrogen should have appeared. But was it intergalactic hydrogen? While an upper limit required only observations at a single frequency, confirmation of a positive result demanded complementary data (the absence of a bump) at another frequency as well. Hydrogen gas emits radiation at a single frequency, 1,420.4 MHz, while other sources of radio energy emit energy over a broad spectrum (the difference between striking a single piano key with your finger and dropping a piano down the stairs).

I would gladly have traded three-quarters of my data for a set of observations from a different frequency to make sure, but it was too late. I wasn't prepared for what turned out to be interfering radiation from a pair of so-called radio galaxies in the same cluster.* Because my single-frequency observations couldn't distinguish between hydrogen and continuous signals from such things as radio emission from exploding stars, my data didn't add much to astronomical knowledge, barely enough to get me a thesis. Was the "empty space" between the galaxies filled with a faint trace of hydrogen? My data didn't answer the question.

Some years later, better observations showed the space be-

*A combination of nuclear energy from exploding stars and electric currents set up by mass motion of ionized interstellar material made these galaxies powerful emitters of radio waves over a continuum of frequencies that included the hydrogen frequency.

tween the galaxies to be devoid of gas—forcing the theorists to change their picture. In the meantime, I finished my studies and took a job at Bell Labs which allowed me to continue my work in radio astronomy.

Consider how my work of 1960 would be done today, in the age of computers. The radio telescopes we now use no longer depend on the earth's rotation. Instead, computer-controlled motors track the astronomical source in question. Every few seconds, the drive moves the beam to a nearby patch of blank sky for comparison purposes. The receiver's data reduction system automatically subtracts the signal at the reference position from the real signal, thereby canceling out the effect of antenna motion and any equipment instabilities. At the same time, the receiver observes several hundred different frequencies simultaneously, each of which is individually recorded and available to the operator via a terminal screen.

Data from previous observations can be retrieved, averaged, combined, and compared with new data. If more than one position is being observed, the computer can provide rough maps in a few seconds. Today, access to on-line data reduction enables each of us to think about our results as we get them. These better tools have profoundly changed the way we work. Today, we can ask questions in time to get answers, make decisions, and create more powerful ideas.

While ideas flow from human minds, computers help to shape much of the information that leads to those ideas. By providing needed information in a timely way and in a digestible form, electronic data-processing equipment can help someone make informed decisions—not just in science, but in most other walks of life as well.

What role should information technology play in your life? I don't think that a lack of technical background need condemn you to polling the "experts." Instead, I hope this book's anecdotes and examples will help you find your own answer to this question.

Information Work

Most of us in the modern world do what is called "information work," a term that covers a lot of territory. An engineer, an air traffic controller, a detective, an economist, a manager, and a stenographer all process information. Decision-making constitutes the common thread that links their work. A decision may be simple—say, selecting a name for a variable in a computer program—or it may be as complex as predicting the future course of the gross national product. Decisions, of course, precede actions, and actions generally result in new information. This rather circular behavior keeps the decision process going until some goal is reached, the task is finished, or the project is set aside for a time.

Complex information work often calls for cooperative efforts. Properly managed, large organizations can tackle complicated tasks far beyond the capability of single individuals—such as flying the Apollo astronauts to the moon and back. But some organizations frequently become bogged down with internal procedures that obscure the original goal and make the exchange of memoranda a self-perpetuating end in itself. In my former hometown of New York City, for example, bureaucratic mixups kept Central Park's ice-skating rink closed to the public for more than seven years before a needed repair job could be completed, accomplished not by government agencies but by the private initiative of an enterprising local real estate developer—a single-minded individual in place of a frozen bureaucracy.

In the modern world, the impact of management decisions extends far beyond corporate boundaries. Consider a community's stake in whether the local Ford plant builds a Taurus or another Edsel. With so much riding on the outcome, an organization's management decisions must reflect the integrated sum of the individual information work done within it. Ford, for example, spent over a billion dollars on engineering before the first Taurus rolled off a production line. In other words, several thousand people spent over ten million total hours "deciding" what kind of car to build. But someone had to decide to start the process and seek the help of others in assembling what proved to be a winning team.

As I see it, a healthy flow of information separates winning organizations from losers. "Deciding" means acting on information. Barring blind luck, the quality of a decision can't be better than the quality of the information behind it. Unlike the president of Ford, who could present the plans for a winning automobile to his board of directors, New York's mayor couldn't extract a viable skating rink repair proposal from the people who worked for him.

The process of moving information from its point of origin to the person who needs it reminds me of an old parlor game called Telephone. The first person in line reads a paragraph and whispers it in the next person's ear. That person whispers an account of what he or she has heard to someone else, and so on down the line, usually with hilarious differences between the original message and the final result. Nowadays, computers get involved in these message-passing chains as well. Furthermore, information often requires processing at its intermediate stages (as when an automobile design moves from styling through engineering to manufacture). It shouldn't surprise us, therefore, that this process doesn't always work—and that a wide gulf separates those organizations whose people and technology do it well from those that do it poorly.

Productive information work can match, quite by itself, many of the economic benefits that flow from natural advantages—such as fertile soil, abundant minerals, or pleasant climate—and has the great additional feature of being totally under human control. Consider Japan and Singapore, two lands with few natural resources but an adequate supply of entrepreneurial individuals who recognize opportunities and take advantage of them in successful enterprises that make productive use of people and technology.

Applications

Success usually comes to those who apply technology to their best advantage. Consider the battle between the telegraph and the Pony Express. For a short time—between April 1860 and October 1861—a small band of brave and resourceful men provided a unique high-speed private mail service. They covered the 1,966 miles between St. Joseph, Missouri, and Sacramento, California, in just under ten days. Changing horses every seventy-five miles, they rode day and night in all kinds of weather, evading hostile Indians, fighting bandits and the terrain, only to be driven out of business by a single strand of copper telegraph wire.

This short-lived Pony Express amounted to little more than a romantic footnote to the history of the American frontier. What fascinates me most about this story is its twentieth-century epilogue. Today, we have a modern counterpart of the Pony Express right in our midst, once again battling the telegraph—only this time the "riders" use jet aircraft instead of horses. An inventive entrepreneur, Frederick Wallace Smith, created the phenomenally successful Federal Express Company, which now provides such fast delivery of documents that many of Western Union's customers have cut back sharply in the number of telegrams they send.

Is speed all? No. The telegraph is a technology that works only with numbers, letters, and other keyboard characters. It is *alphanumeric* and, of course, the alphanumeric format of telegrams limits the user to text alone and requires word-by-word retyping in order to enter text into the system—a cumbersome and error-prone process. In contrast, Smith's latter-day Pony Express can transport anything at all.

The Federal Express Company is an "information-work" enterprise that has combined advanced technology with a powerfully simple concept, to provide overnight package delivery service between almost any two points in the continental United States. The heart of the system depends on collecting incoming mail at a single, centrally located jetport (in Memphis) each evening, sorting it overnight, and delivering it nationwide the next morning.

Reliability of service is assured by sophisticated equipment-scheduling and package-tracking systems. For example, package deliverers carry pocket-sized data entry devices with them. Just before a package is handed to its intended recipient, anywhere in the United States, the deliverer uses the device to scan the bar code on the package, entering the recipient's name and location. On returning to the truck, the deliverer transmits that information to a central tracking computer in Memphis. It's a truly high-tech operation.

While efficient use of technology appears to have carried the latter-day Pony Express to a clear lead over the telegraph, winners cannot rest on their laurels in a high-tech world. A new transmission technology is already gaining momentum. Facsimile machines—table-top "copiers" that can send graphical replicas of letter-sized documents over existing telecommunications facilities—now offer telegraphic speed without the need for retyping. With hundreds of thousands of these machines being installed every year, overnight delivery of documents will surely become less attractive. In all areas of business, the changes brought about by new technology will probably leave some of today's leaders scrambling to catch up to more nimble competitors.

An ancient Chinese curse says, "May you live in interesting times." As a technology *provider,* my life is certainly "interesting," working in a company that races to give its customers the benefits of the latest advances in technology. But technological change doesn't always lead to changed life-styles. Many technology *users* with whom I come in contact don't always employ the advanced technology they buy to change the way they do things. Quite the contrary. People in many walks of life take new technology very much in stride.

Consider the computing that an automobile accident can call for: a microprocessor (computer on a chip) analyzing the driver's breath for alcohol level; computer controlled lab instruments scrutinizing forensic material; reporters hunting through computerized information sources for background information; lawyers matching the issues against the legal precedents found by key-word search programs; insurance company computers handling claim data.

This amounts to a vast network of computerized behavior. Despite all this computerized help, however, the cast of characters in this hypothetical automobile accident plays much the same set of roles they might have played in the 1930s. Today's police, reporters, lawyers, and insurance agents carry out their jobs in ways their grandparents would have recognized. In each of these specialties, computers act mainly to strengthen existing human capabilities—supplying different modes of operation—without changing the underlying job structure.

Such nonrevolutionary changes typify a large, and I think underappreciated, segment of the "information revolution," applications in which people adapt computers to do their existing jobs more effectively and conveniently, rather than to do them in a fundamentally different way. For all the talk about the impact of computer revolution technology, most people haven't experienced revolutionary changes in their lives—just because computers have allowed them to keep pace with today's faster world. In most cases, the presence of computers gets far less notice than their occasional absence, as when a computerized reservation system goes down at a busy airport.

As I see it, many significant effects of information technology escape general notice because they act primarily to cushion us against impending disruption, rather than to bring about change. This was not true in the nineteenth and early twentieth centuries, when the technology that drove the industrial revolution served primarily to increase the production of material goods. Today's "post-industrial" society instead devotes much effort to maintaining the quality of life in the face of insufficient human and material resources—ranging from a shortage of land and minerals to an underskilled labor pool.

The story of the New York jetport that never got built exemplifies the underappreciated use of information technology as a resource cushion. In 1959, the Port Authority of New York and New Jersey attacked the increasing traffic load on its La-Guardia, Kennedy (then Idlewild), and Newark airports by calling for the construction of a fourth airport in an "unused" area, Morris County's Great Swamp in central New Jersey. An immediate uproar ensued as New Jerseyans fought to prevent the bulldozing of this unique wildlife area. After a long and hard-fought battle, the Port Authority retreated to try its luck elsewhere.

Airport planners next zeroed in on a little-used air force base in Newburgh, New York. Prospects brightened as local leaders in this rather depressed area signaled a welcome to this boost to their economy. But it was not to be. The airport never got beyond the public relations stage, for quite abruptly the planners reversed themselves. There was no need for a new airport after all.

What had happened? The need for greater capacity had driven the aviation industry and the federal government to create more efficient means of dealing with air traffic. Better radar tracking and flight control procedures, as well as the then-new wide-body aircraft design, yielded enough extra capacity at New York's three existing airports to eliminate the need for a fourth. Thus, the application of information technology created the capacity of a multibillion-dollar airport without the side effects. People didn't take much notice of that event because cushioning New York from the consequences of air traffic growth didn't make the headlines.

The technology-based solution to the airline capacity problem proved adequate to handle far more traffic than originally planned. Not until U.S. government deregulation of the airline industry precipitated a massive increase in air traffic in the mid-1980s did the capacity problem rear its head again. As of mid-1988, federal officials are preparing once more to increase airport capacities with better aircraft-tracking technologies. For example, normal aircraft landing procedures depend on visual contact between planes to maintain their positions in curved flight paths. In periods of poor visibility, pilots are forced to land via less efficient straight-line paths by "riding" on the beam of a radio beacon. A proposed new radar system would eliminate the need for visual contact. Until then, however, airline schedulers must either pray for continuing sunshine or find extra airports.

Modern automobiles provide a second example of underappreciated computer impact. The cars we drive today helped us survive the oil crisis of the 1970s by making millions of barrels of imported oil unnecessary. We need a lot less petroleum now because today's automobiles burn fuel more efficiently than their predecessors. My new car, for instance, works about as well as my old gas guzzler did but it doesn't consume nearly as much gas, thanks to the application of information technology—such as computer-aided design of lightweight structures and efficient drive trains, as well as the on-board microprocessors that work to minimize fuel consumption.

In a way, automobile designers have created petroleum out of pure information.* This was another drama muted because the story became a "non-event" as the focus of our attention—the oil shortages—disappeared. Ponder those millions of all but unnoticed computers humming along under millions of automobile hoods. Combined with similar energy savings in other applications, they took the edge off the oil shortage, and OPEC's control crumbled.

On a smaller scale, I've found cases where technology has compensated for unwanted changes on a personal level. The neighborhood stores I used to patronize have been replaced by look-alike shopping malls. Still, I can get the special things I want thanks to mail-order catalogues, 800 numbers, credit cards, and the specialized computing needed for successful telemarketing. One such computer system makes my favorite source of outdoor clothing a pleasure to patronize. When I spot something I like in their catalogue and call them, the clerk checks the inventory on a computer screen to make sure that the item I want is in stock—in my size and color—earmarks it for me, and arranges for shipment and billing, all with a few keystrokes. If I call in the morning, the package sometimes arrives the next day. Smooth and simple.

On the other hand, technology, as we all know, often works fitfully if at all. I sculpt as a serious hobby, and several years ago I was in need of some gears to complete a motor-driven kinetic sculpture I was building with my friend Lillian Schwartz—the same artist who did the computer graphic on the jacket of this

---

*Some of this improvement could have been accomplished with pencil and paper if computers hadn't existed. The story concerns how people used technology, rather than the properties of the technology itself.

book. We had committed ourselves to an out-of-town showing; time was short. Fortunately, I remembered having used similar gears in an astronomical clock drive back in the mid-sixties and I was even able to dig out the old catalogue from the folder where I had stuffed the old drawings. The phone number on the catalogue cover still worked, as did the part numbers inside. Except for a steep increase in prices, nothing seemed to have changed. I quickly picked out the parts I wanted and proceeded to order them via my personal credit card.

Delighted at my good fortune, I thanked the young woman on the other end and told her how relieved I was to know that the gears would be on their way that same afternoon. Then the blow fell. "We can't ship them till next week," she said. "What?" I sputtered. "It says right here on the cover of your catalogue that orders received before 11 A.M. are shipped the same day." "You must have an old catalogue," she replied, without the slightest hint of irony in her voice. "Now we have a computer."

So, in a world of "information work," we seem bound to this thing called a computer, a machine that helps some to move ahead, others to remain in their accustomed places, and still others to move backwards.

Computers

And just what is this thing that structures so much of our lives? Fortunately, you don't need technical training to gain a meaningful grasp of its inner workings. The deeper one goes into the principles that underlie computers, the simpler these machines are. The intimidating arrays of control keys and cryptic commands need concern only those who operate them. Instead of troubling you with this complicating surface layer—which will change anyway as technology evolves—I will instead focus here on the underlying operations you need to understand in order to judge how they can help you.

Stripped of its interfaces, a bare computer boils down to little more than a pocket calculator that can push its own buttons and remember what it has done. Obeying a stored sequence (or *program*) of instructions, it moves, adds, subtracts, multiplies, and divides numbers at blinding speeds, hauling them in and out of its memory as needed. Individually, each of these operations would require in human hands nothing more than paper and pencil. Computers merely add speed and diligence to the process. Most of a computer's operations come straight from mathematics (like addition and multiplication), while others come from logic (such as, "If A equals B, skip the next instruction").

Mathematics is among the most explicit (or judgment-free) of human cognitive processes. Its mechanization predates even the invention of the abacus in ancient times, and forms the basis of modern computing. Judgment, on the other hand, requires the exercise of human intelligence.

When computers appear to apply judgment, they are actually following human instructions embodied in the programming process. For example, a computer might "decide" that a business ought to lease trucks and hire drivers, instead of working through an independent trucking firm. In reality, however, the computer has only decided which of two sets of numbers has the larger sum. The programmer's judgment* connects each of the relevant pieces of the problem with a corresponding number, either by putting a price on each item directly or by establishing a procedure for looking up tabulated data (like the consumer price index, or the "prime" rate for business loans).

Computers offer us vital help in clarifying the quantitative aspects of problems. They provide us with a powerful mechanical extension of certain human data-processing skills, notably

*Or the judgment of some other person upon whom the programmer depended for expertise.

math and memory, but no help at all with others, like artistic inspiration. Michelangelo's magnificent statue of David provides a beautiful example. Michelangelo "saw" the young shepherd, leaning naturally on one foot and balancing himself with an extended elbow, within an oddly shaped piece of marble which lesser sculptors had rejected as too difficult to work with. That stone didn't fit the formal rules of balance and composition, yet from it Michelangelo created one of the world's greatest masterpieces.

While Michelangelo's work *inspired* countless artists, he couldn't *instruct* others to create comparable masterpieces. Artistic genius still defies codification.* On the other hand, once the first Western mathematician learned how to do long division, it was only a matter of time before that capability spread to every corner of Europe. After all, people can only instruct others in what they themselves know how to do at a conscious level, the explicit (as opposed to the intuitive) portions of human cognition. Explicit processes (like playing tic-tac-toe or doing long division) transfer easily from one person to another or from people to machines. Just write out the instructions. But how do you instruct someone (or some computer) to recognize caricatures in a magazine, let alone find a human figure in a misshapen piece of rock? In other words, how do you write the software? Some of the brightest people I know are working hard to solve such problems—with only limited success so far.

Since computers all operate according to the same laws of mathematics and logic, software rather than hardware most often limits the range of problems that lend themselves to computerization. Computers are most effective in dealing with the

---

*While theory helps us appreciate great works of art, the mechanistic application of its rules rarely leads to masterpieces. Consider Rome's Victor Emanuel Monument. Its designers followed all the rules but produced a huge architectural joke.

kinds of data that translate readily into numbers. As a result, explicitly structured problems that lend themselves to "by-the-number" solutions give programmers the fewest headaches. For example, if the students in a high school computing class were assigned the creation of a tic-tac-toe program as a homework assignment, most of them would probably be able to write one. On the other hand, a program that could distinguish between male and female faces in random snapshots would probably earn its author a Ph.D. in computer science. Small wonder, then, that most programs focus on the numerical aspects of problems.

While a well-maintained computer normally handles its numerical tasks with unfailing accuracy, a machine can't assume responsibility for the relevance of those calculations to the real-world problem at hand. Take the "meltdown" of the New York Stock Exchange on October 19, 1987, for example. The market lost more than 20 percent of its value in a single day, with the Dow Jones average dropping more than five hundred points. Several large financial institutions bet huge sums of money on their computers' abilities to make split-second buy-and-sell decisions on the basis of price behavior. These systems traded shares, options, and futures against one another faster than a human trader could follow. Until that morning, such methods seemed foolproof. As that day proceeded, however, buyers failed to appear for the "sell" orders the machines had issued. Transactions failed to execute as some market makers stopped answering their phones—situations that the creators of these trading programs hadn't anticipated. The experience proved a costly lesson for all concerned. Simple numerical models rarely represent reality in all its aspects.

The October meltdown was simply old poison in a new bottle since oversimplification and rigidity of inputs to decisions greatly predate computers. Throughout history, celebrated "managers" fell victim to the temptation of letting numbers displace judgment. Napoleon Bonaparte's efforts to manage his navy is a good example. Frustrated by the apparent cowardice of his admirals, he decreed that his fleet attack the enemy whenever the French had more ships in an area than the British. This was a fatal edict.

On October 20, 1805, Napoleon's admiral Pierre de Villeneuve sailed his thirty-three ships from the safety of Cadiz Harbor to challenge Horatio Nelson's battle-ready squadron of twenty-seven ships. Clearly, Villeneuve held the upper hand numerically. Just as clearly, Napoleon's orders forced the decision. Not so clear was the firepower advantage held by English gun crews—drilled to fire and reload far more quickly than their opponents—and Nelson's cool nerve in leading his squadrons straight through the enemy line. By the evening of the 21st, Napoleon's hopes of sea power lay smashed forever on the shores of Cape Trafalgar.

That fateful event hinged on a numerical calculation, a calculation whose outcome had nothing to do with the mechanism used to execute it. Villeneuve's decision would have been the same had he counted with his fingers, used pencil and paper, or worked an abacus. The issue hinged instead on Napoleon's one-line "program" and the mistaken assumption that went into it—that warships were alike no matter who manned them or who led them.

Upon reading this account, my friend Jon Bentley suggested that I point out possible improvements to the program, such as multiplying the number of guns on each ship by the maximum rate at which each crew could fire them. An interesting exercise. First of all, British gun crews were drilled to fire about twice as a fast as their French opponents. Furthermore, French doctrine called for firing upward into the enemy's rigging in the hopes of crippling its ability to maneuver, while the British fired their broadsides into their opponent's hull. This steady rain of shot (and the lethal shower of splinters which multiplied the effect of each cannonball) created havoc on the French gun decks and reduced their firepower.

Had Napoleon understood these factors well enough to put them into his program, he would have realized that he needed a change in doctrine and training, rather than a computer, to help direct his navy. In that way, the French might have done as well as the newly formed United States Navy, whose crews were trained on the British model. Although the British Navy dominated U.S. coastal waters during the War of 1812, it did so only by sheer weight of numbers. Man-for-man and gun-for-gun, the tiny American republic's warships proved every bit as effective in battle as their British opponents.

My point is this: while conventional wisdom often tends to blame "computers" for foul-ups, such results rarely have much to do with the mechanism that executes the underlying mathematics or logic. Instead, the credit or blame should go to the people who fail to understand the problem they assign to some particular information-processing system.

Contrast this misuse of numbers with unexpected benefits in other applications. While overreliance on "numbers" has gotten some problem solvers into trouble, assigning numbers to other entities has yielded fantastic dividends. Almost everything seems "digital" today. Just look at today's home appliances. It's hard to find one that still has knobs to turn. When you want to select a TV channel or set the timer on your microwave oven, you just punch in the digits and the circuits do the rest electronically.

This digitization of consumer electronics is just a small part of a much larger trend. The singular success of mathematical computing has displaced the need for the direct (or analog) mechanization of other processes. Instead, designers convert all sorts of tasks, ranging from word processing to robotic Ping-

Pong, into their equivalent mathematical forms—letting a computer do the job rather than attempting to build specialized machines. As long as a process lends itself to explicit description, our programming methods can usually provide a means of creating its computable equivalent. Furthermore, digital technology has moved out of the computer field and into a wide variety of other applications, such as modern telecommunications.

This technological change has forced most of us to unlearn much of what we once "knew." Shortly after joining Bell Labs' Radio Research Laboratory in 1961, I thought I had uncovered a new idea for reducing the cost of telephony. My idea was to carry a very large number of telephone circuits* over a single high-capacity microwave radio link. With many circuits sharing the same transmitter, receiver, and antennas, the cost per circuit would become infinitesimally small.

Unfortunately, my colleagues had heard it all before. In fact, most inexperienced radio engineers "discovered" that idea early in their careers. Quite to my dismay, I soon found my error, which was a matter of the *channel units*—the devices that merge telephone circuits together on one end of a common channel and redivide them on the other end, a little like the switchyards railroads use to assemble boxcars into freight trains. In those days every merging of telephone circuits in a shared transmission facility required such devices. Moreover, they were very expensive, as an experienced team of engineers had worked out—each pair of wires needed its own channel, which meant $600 per line, no exceptions. So, if I had a system for simultaneously transmitting a thousand telephone circuits, the channel units alone would cost $600,000—enough money to string a lot of individual wires instead. Furthermore, every time my system came to a switchpoint in the network, it would

*Each "circuit" represents a connection between the two ends of a telephone conversation.

take an extra set of channel units to route each of the individual calls through that switch.

All conceivable ways of lowering the channel units' price had already been exhausted, I found on examining one. Each vacuum tube had been replaced with a transistor. The need for each coil, capacitor, and resistor had been examined, and each had been found to play an indispensable role in shaping and stacking the signals together. The most expensive remaining part in each unit was a transformer that cost just $8. Furthermore, it cost almost $200 to provide all the insulated connecting screws each unit needed to attach the various kinds of wire pairs the system might encounter. So much for Penzias's revolution in communications!

Once digital electronics came along, however, the entire picture changed. We still use channel units to combine and break out individual telephone calls, but they do much more and cost much less. The old individual transistors with their coils and capacitors are gone, replaced by little chips of silicon upon which thousands of transistors work together to process the signal. All but one pair of the old connecting screws are gone, too, because the circuitry can adapt to any input on command.

The input circuit converts the caller's voice into digits. These digits move through the network in high-capacity systems that can swap individual calls without sorting them out first. Instead of the old method of connecting one pair of wires to another, a mathematical switching operation does the job.

The science I learned in the sixties is still true. The laws of nature that govern telecommunication haven't changed, but digital technology has given us a loophole that makes those laws almost irrelevant. Today, engineers contemplate shipping a quarter of a million simultaneous telephone calls on a single strand of optical fiber, with room left over for data and even video.

Even more important than this dramatic increase in communications capacity is that the digitization (conversion into numerical form) of a signal offers us the opportunity to apply data-processing techniques in a variety of applications. These can range from simple numerical addition when combining several participants in a conference telephone call, all the way to computerized language understanding. Though still in its infancy, this latter capability enables people literally to talk to machines that "understand" them. Some computers have been programmed to compare the digitized signals derived from human speech against stored examples. Each stored signal corresponds to a particular phonetic sound, so a match tells the computer which sound it has "heard."

That ability to match sound sequences against the sounds of common English words gives such systems a rudimentary "vocabulary." As the art of creating sentence-parsing programs advances, we can look forward to machines that can extract meaning from ever more complex commands spoken in everyday English.

Computerized understanding of language is just one example of the vast opportunities for new applications that such technology offers us. As I have tried to show through the anecdotes and examples, however, technology can lead to pitfalls for the unwary. Simple self-interest, therefore, demands an understanding of technology and its impacts.

I believe that a meaningful personal assessment of how information-processing technology can and ought to shape our lives must begin with an understanding of information itself and what it means to "process" that substance. What is this raw material that people and their machines use to make decisions? Accordingly, the next chapter will consider information in its most basic form—the common symbols.

Chapter 2

Numbers, Words, and Pictures

A book is a machine to think with.
—I. A. RICHARDS, *Principles of Literary Criticism*

ALL USES of information depend on symbols. For example, the act of reading a book brings an arrangement of symbols to the reader's attention—words, numbers, pictures, and punctuation marks, as well as specialized items, such as mathematical operators, chemical formulas, and musical notes. Symbols stand for ideas, conditions, qualities, and other abstractions. Individually and in groups, then, they constitute the raw material people and computers use to get information. By beginning this examination of information handling with a look at the symbols themselves, I hope to lay the groundwork for your firsthand picture of the benefits and pitfalls of using computers to deal with various kinds of information.

Each living brain creates and manipulates its own set of symbols as it acquires, processes, stores, and acts upon sensory information. But until recently, we humans knew very little about the symbols our brains create, certainly not enough to use them as a model. In fact, words, numbers, and pictures were all invented long before people even knew for sure that such a thing as the brain existed. As a result, our ancestors couldn't have modeled their symbol systems after nature's even if they had wanted to. Instead, humanity created symbol systems out of convenient objects, such as raised fingers for counting, or rhythmic drumbeats for communication. The modern descendants of these ad hoc arrangements act in partnership with the human brain but don't necessarily work in the same way any more than a high-fidelity stereo uses a set of miniature instruments to play symphonic music.

The fundamental difference between the way a sketchpad and a group of brain cells represent the essential features of a cocker spaniel, say, would be of little more than academic interest if the human brain remained our only information-processing tool. But we now have another. Computers now offer us an alternative way of dealing with many kinds of information—thanks to programming methods which recast the symbols that carry such information into a form that such machines can handle.

Human beings can learn to identify flowers, speak Portugese, or diagnose chicken pox without concerning themselves about where and how their brains will store the information. Computers, on the other hand, deal primarily with numerical information. As a result, early computer scientists had to find ways of casting non-numerical information—such as text and images—into equivalent forms that lend themselves to computer processing. For example, the designer of a chess-playing program normally assigns a numerical value to each of the squares on the board, so that the program can keep track of piece positions.

Symbols are tools. Just as levers and wheels enhanced our ancestors' ability to move objects, number systems and writing enhanced their ability to solve problems. Before we examine the most important kinds of symbols and their relation to computing, I'll introduce the subject from a historical perspective.

History

About ten thousand years ago, the first pictures appeared in the magnificent caves of France and Spain: not all at once, but the crudely traced outlines of human hands placed against the rock, followed by drawings of bison, horses, and even woolly rhinoceroses. This was a huge leap in consciousness, for once people had connected intentional marks on a surface with a familiar object, pictures could be understood to stand for objects in much the same way as words.

We can imagine these early artists pointing to their pictures and saying "bison" or "horse." Naming—using a word to stand for something else—comes naturally. The next step was harder and took a long time coming. Not for many thousands of years of picture making did someone make the inverse connection, using "something else" (a figure) to stand for a word. Figures that stand for words define written language.

The symbols that convey the information contained in this book owe their existence to a series of remarkable inventions. Like many other conveniences of modern life, the deceptive simplicity of our phonetic alphabet and decimal numbering systems makes them appear "obvious" in retrospect. But history tells a different story. Progress was contingent upon an uncertain chain of events.

The alphabet and the decimal system spread from the Middle East and India to all of humankind. But news of these discoveries moved haphazardly across the globe. By the time the decimal system reached France in the middle of the twelfth century, for example, the University of Paris had been in operation for over three hundred years. Mathematically speaking, medieval Europe was an intellectual backwater. While few modern builders would attempt even a small chicken coop without first doing a few calculations, the creators of Notre Dame could barely multiply two small numbers together, and had no means of doing long division.

Furthermore, while mathematical knowledge in medieval Europe lagged far behind that in Asia and North Africa, a few parts of the world had no knowledge of mathematics at all. In fact, at least one society—New Guinea's Gimi tribe—survived well into the twentieth century without knowing that the art of counting even existed!

Until the Gimi tribe made first contact with modern outsiders in the aftermath of World War II, they had no numerical quantifiers beyond "one" *(iya)* and "pair" *(rarido)*. When anthropologist Jared Diamond asked them to quantify a group of seven similar objects, for example, they would reply *"iya, rarido, rarido, rarido."* Since they lived by hunting and gathering food, they encountered few situations that called for numeration. When obliged to deal with such problems, they employed the tallying techniques first used in the Stone Age—making a series of scratches on a piece of wood for each of the objects, people, or events being tracked. (We moderns sometimes resort to tallying for special tasks such as determining the results of an election. As the votes on each ballot are read out loud, someone makes a corresponding series of slashes under the names of the candidates—often slanting each fifth mark across the previous four to simplify the subsequent counting operation.)

The earliest archeological examples of tallying—multiple notches cut into animal bones—date back thirty thousand years. A variety of other techniques and materials for the conduct of life have come to light in excavations of more recent origin, but no clues to their creator's counting (as opposed to tallying) abilities survived. Since human beings of this period lacked any permanent form of expressing these concepts beyond the marks themselves, we can only guess at the kinds

of objects they represented, and the degree to which these primitive accountants grasped what they were actually doing.

This picture changed dramatically with the advent of a new
kind of recordkeeping—the clay tokens of Mesopotamia. About
ten thousand years ago, the inhabitants of that region (centered
upon modern Iraq) began using distinctively shaped clay tokens—spheres, disks, cones, cylinders, and triangles, among others—to keep track of foodstuffs, livestock, and land. Thus a
merchant might have tracked baskets of grain as they moved
in and out of his warehouse by adding to and subtracting from
a pile of spherical bits of clay. Thanks to the work of archeolo-
gist Denise Schmandt-Besserat, who deciphered them, such
tokens help trace the rich variety of goods that flowed through
commerce in that cradle of civilization.

Interestingly, these tokens appeared at a unique moment in
human history, the advent of agriculture. The ability to culti-
vate food, instead of constantly searching for it, provided the
foundation for all the human progress that followed. As life
became less subject to the whims of nature, it also became more
complex. People began to own, to store, and to trade more
things, so they naturally needed far better ways of keeping
records. Fortunately, someone figured out the token scheme.
As a result, this great technological advance also brought about
a dramatic improvement in the ability to store information
outside of the human brain.

The token system remained in use in the Middle East for
almost five thousand years without significant change, other
than a large increase in the number of shapes used—from some
two dozen at the start to over two hundred.* Then, around
3300 B.C., the inhabitants of Sumer in present-day Iraq adopted
the practice of storing tokens in sealed clay jars in order to keep

*As the system grew more complex, the shapes even included miniature
models of tools and animals.

them together and thereby create more permanent records. Unfortunately, however, the only way to examine one of these records was to break open the jar.

At this point, ingenuity intervened once again. Someone hit upon the idea of making a series of marks in the soft clay cover as the jar was being sealed—one for each of the tokens that the jar contained. For example, a long, wedge-shaped incision indicated that the jar contained a cylindrical token. That token, in turn, represented one goat. Since most jars contained appreciable numbers of the same tokens, the Sumerians soon adopted the practice of using the symbols on the jars for grain measure to denote multiple instances of other tokens. For example, a hemispherical impression,* the symbol for a *bariga* (about a bushel) of grain, next to the symbol for the goat token indicated that the jar contained ten goat tokens. A slash separated these two marks in order to distinguish this situation from a jar containing a grain token and a single goat token.

But once the markings had captured all the information about the jar's contents, the contents themselves became unnecessary. Archeological evidence shows that the marked jars led almost immediately to a system of marks on clay tablets. These rudimentary numbers and pictographs provided the foundation for cuneiform, the first written language.

While the Sumerians were clearly first, the idea of using pictographs to represent words also developed in at least four other places around the world. Hieroglyphics appeared shortly afterward in nearby Egypt. Since no hard evidence of a link between the two events has come to light, most scholars regard them as independent discoveries. India and China also developed systems of their own, as did the Mayans of Central America.

---

*Which, in this case, corresponded to a spherical token.

The Mayan system proved the shortest-lived. Invented just a few hundred years before the Spanish conquest of the region, it was soon supplanted by European writing methods. At the other end of the scale, the Chinese alphabet survives to the present day and remains largely intact.

One reason for the durability of the Chinese alphabet stems from the monosyllabic nature of the spoken languages that employed it.* Since each character corresponded to a syllable, the language could readily accommodate foreign words and names. For example, I served under an army officer named Federen who owned a small rubber stamp with the Chinese equivalent of his name on it, "foo-deh-ho"—literally "love," "virtue," and "peace"—which was a souvenir of a tour of duty on Formosa.

By contrast, Egyptian hieroglyphics proved far less suited to forming foreign words and names than the Chinese, because their spoken language consisted mainly of polysyllabic words. As a result, the Egyptians adopted a system based on the beginning sounds of common words. In this scheme, a mark in front of a word would tell the reader to voice only the leading syllable, so the word sequence "*feather, *dentist, *handkerchief" would be pronounced "fedeha." For unknown reasons, the Egyptians refrained from extending this system to their own language as a whole, but their eastern neighbors had no such scruples.

The Midianites of the Sinai readily adopted for all of their language the shortcut that the abbreviated set of characters offered. By that time, the system had been refined to some

---

*For an account of the problems caused by the use of Chinese characters with the polysyllabic Japanese language, I invite interested readers to see J. Marshall Unger's *The Fifth Generation Fallacy* (Oxford University Press, New York, 1987).

two dozen consonants—much like modern-day Hebrew. Soon thereafter, neighboring peoples adopted it and spread it eastward all the way to India and westward throughout the Mediterranean—most notably to Greece where vowels were added and the basis for all Western alphabets was established some twenty-five hundred years ago.

The invention of the Greek alphabet set the stage for the explosion of knowledge that took place around the shores of the Mediterranean. As we look back across the centuries that separate that unique age of Greek creativity from our own, we try to imagine how such a tiny handful of humanity managed to accomplish all that it did. The staggering achievements of Greece do credit to all mankind, but Athens at its peak was a small place with barely enough inhabitants to rate a ZIP code in our modern society.

So much was created then—religious teachings, philosophy, logic, geometry, science, all kinds of art forms—that for almost two thousand years most of "educated" mankind felt it had little more to do than study and replicate these achievements of the ancient Greeks. Ironically, numbers and numerical concepts largely escaped Greek interest. The educated classes regarded numbers as more suited to the needs of commerce than to the kind of theoretical study they lavished on geometry, for example. As a result, our modern system of numbers came to us through other people.

Just as pictographic writing preceded the alphabet, less sophisticated numerical symbols provided the needed preliminary steps on the way to our modern decimal system. As we have seen, primitive people found a way of dealing with numerical information thousands of years before anyone learned how to count. They relied on tallying instead. Imagine having to entrust a small flock of animals to the care of a shepherd who

couldn't count. A simple set of notches on a stick could solve
that problem. Run one finger down the notches one by one
while pointing to each of the animals with another.

The use of fingers in tallying naturally led to organizing tal-
lies into groups of five and ten, and subsequently to the widespread use of those numbers in later numbering systems. As
people learned to count, they continued to make extensive use
of their fingers. The Babylonians counted from 1 to 12 by running the tip of a thumb horizontally over the inside segments
of the fingers on the same hand. They gave each of these segments a name, so that "middle joint of the ring finger" might
have stood for 6.

For larger numbers, a Babylonian could keep track of repeated counts on one hand by folding over a finger of the other
hand every time the thumb reached 12. With five repetitions,
they could conveniently count to 60. That number became the
base of the Babylonian numbering system, and remains embed-
ded in timekeeping to the present day. The conventional division of hours and minutes into sixty parts owes its origins to the
number system used by the people who taught the rest of
humankind how to measure the passing of time.

For writing purposes, however, the base-60 system was simply too awkward to survive for long, and most civilizations
sooner or later adopted base-10 systems. For example, the
Egyptians had distinct number symbols for 1, 10, 100, 1,000,
10,000, and 100,000. The Romans "improved" upon the Egyptian system by adding extra symbols and combinations of symbols to reduce the amount of writing needed. Thus 1964
becomes MCMLXIV in Roman numerals.

Had the Romans merely adopted the Egyptian system, the
same number would read MCCCCCCCCCXXXXXXIIII.
While MCMLXIV is easier to write, it makes arithmetic really

awkward. Nevertheless, the Western world struggled along with this system until the Norman conquest of Saracen-held Sicily in 1091. Fortunately, the conquerors recognized the value of the manuscripts included in their booty. As a result, Europeans gained access to the place-notated decimal system that forms the basis of modern mathematics. While Westerners received this system from the "Saracen" Arabs, "Arabic" numerals depended in turn on Indian inventions—most notably the zero—which make convenient arithmetic possible.

The science and technology of twelfth-century Europe lagged far behind that of the more advanced East. But not for long. As the influx of new knowledge began to pull Europe out of the Dark Ages, access to mathematical tools made possible advances in other areas. For example, new methods of calculation permitted closer examination of the motion of the heavenly bodies and led directly to the astronomical revolution brought on by Copernicus, Kepler, Galileo, and Newton. And so into the modern world, where mathematics underlies every facet of life, from the aerodynamic calculations that shape the curved outlines of a jetliner to buying groceries and planning dinner. From a state of almost total ignorance we have moved, then, to a state of almost total dependence on numbers. For most of us, life without numbers would be unthinkable. Clearly, these pervasive symbols deserve our attention.

Numbers

Modern mathematics grew out of a combination of invention and discovery. While people *invented* the decimal system and the multiplication method that we all learned in school, they *discovered* the fact that the product of 3 times 5 yields exactly the same result as 5 times 3. That's true no matter what multiplication method one devises. And 7 has always been a prime number,* no matter what symbol people used to represent it, or whether or not they even knew that numbers existed.

The mathematical laws that govern the behavior of numbers exist in their own world, independent of any physical reality. For example, one can count *without counting anything,* real or imaginary. Because these mathematical laws also apply to key aspects of the physical world, human beings have been able to explore these principles and put the results to advantageous uses.

Unlike the products of human invention, the laws of arithmetic existed long before our ancestors discovered the handy correspondence between one or more raised fingers and an equal number of objects. Naturally, human beings first learned about math by experimenting. Two fingers plus three more fingers always equaled five fingers. As humans found more and more objects exhibiting this same property, they extrapolated these observations to create the abstract notion of numbers and counting that laid the foundation for understanding mathematics. (The discovery of the number *zero* provides a fine example of such an abstraction.)

It is logical to assume that when children learn math they recapitulate much of the experimental process by which our early ancestors gained their first notions of this subject. "If Julia has five apples and gives two to William, how many apples does she have left?" Good teachers illustrate arithmetic with familiar objects and invite children to carry out experiments, or at least imagine them. This is real life. People of all ages grasp concrete examples more easily than bare abstractions. Having learned to subtract apples, they can go on to apply the same method to similar situations involving other objects.

*Division of a prime number by any other number (except the number *one*) leaves a remainder, that is, a fraction.

It follows that the role of context in arithmetical problem solving serves to distinguish human information processing from the way computers operate. Human intelligence almost always thrives on context while computers work on abstract numbers alone. A subtraction problem cast in terms of apples is easier for a student, but a programmer must do precisely the opposite thing, convert concrete problems into context-free mathematics, because that is easier for the machine. If I had to create a program to keep track of fruit, for example, I would arrange to have the fruit name (attached to each number) stripped off and stored on the way in, then reattached to the answer on the way out, much as we Northerners store our overcoats in a restaurant.

Independence from context is in fact a great strength of mathematics. Once some portion of a problem has been cast in numerical terms, all the power of computation becomes available to attack it. As a result, an airline and a supermarket chain can use some of the same mathematical recipes, or algorithms—one to schedule its aircraft and the other to allocate space on its shelves. Numbers attach themselves to 747s and Twinkies with equal ease.

Say American Airlines has eighty-four gates at its Atlanta hub, and my local Foodtown has eighty-four feet of aisle space for baked goods. One company needs to estimate available gate time, the other shelf space. In the former case, the computer would multiply the number of each type of aircraft by its turn-around time, add the results together, and test to make sure that the sum did not exceed the product of the number of gates times the number of hours the airport is in operation. In the latter case, the computer would multiply the width of each item by the number of rows allocated to it, add those results together, and test to make sure that this sum didn't exceed the product of the number of aisle feet times the number of shelves in the cabinets. Quite different contexts, but the math produced by both these approximations is identical.

Contrary to the popular view, computers sometimes make mathematical mistakes, but by far the more serious source of potential error lies in the interface between the computation and the real-world problem to which it applies. A mathematically correct solution may work fine in one context but fail in another. In the previous chapter, I recounted Napoleon's disastrous use of warship counts as the inflexible measure of naval power, a simple case. Most of today's abuses of numbers are far harder to spot because they lie deeper in the problem, as the following business story demonstrates.

In the mid-1970s, Aamco, a leading wholesaler of automobile transmission parts, decided to diversify by opening a chain of gift boutiques called Plum Tree. The plan called for the same system of centralized buying and centralized reselling to franchise that had proven profitable in the transmission-repair business. Buying in volume cost less, and franchises could buy individual items as needed. By tying central purchasing to a careful monitoring of each item's sales, the system kept inventories low while still maintaining the levels of stock needed to support the sales effort.

The program began promisingly as the first of the Plum Tree stores opened. Customers responded favorably to the array of gift items presented and bought enough to justify further expansion. So far so good. Then came the bad news, when early promise faded as Plum Tree sales topped out and then declined—declined so far that management decided to quit that business and look for brighter prospects elsewhere.

What had gone wrong? A cultural mistake had been made. The ordering and inventory control system that served so well in the transmission business actually worked against Aamco's interests in the gift boutique business. The purchasing system kept track of the numbers and prices of canisters, dish towels, trays, and soap with the same objective precision it had always given to bearings, washers, and gaskets. Since the two sets of items shared the same numerical properties, mathematics produced the same results regardless of which items the numbers represented.

The system automatically ordered larger quantities of the most frequently used transmission parts in order to maintain the flow into the repair shops, and to achieve a more competitive price on large-volume items. This same system also reordered more of gift items the Plum Trees' customers purchased most frequently. But automobiles and boutiques are not the same. A mechanic will usually repeat the same sort of repair day after day and therefore will need the same parts repeatedly; boutique customers, on the other hand, rarely buy the same gift for their friends time after time. Presented only with more of what they had already browsed over, the customers stopped coming.

The lesson: an attractive "objectivity" of numbers emphasizes similarities. The human user must perceive and allow for whatever differences exist in the real-world entities that these objective symbols represent.

But even when they are properly used, numbers can aid only in the solution of those problems—or portions of problems—that lend themselves to numerical treatment. To bring computing to bear upon a richer variety of problems, computers need the ability to recognize, store, and process non-numerical information, notably in the form of language and graphics. As a result, much of the utility of present-day machines depends on their ability to deal with words and pictures. Let's see how well they are doing.

Words

As symbols, words offer us a convenient format for mechanization as well as an expressive way of representing contextual
knowledge through the use of language. Modern computers
accept the letters of the alphabet as readily as numerical inputs.
An electronic interface automatically converts each keyboard
character into an equivalent number according to a standard
format. It's all numbers to the machine. The computer stores
these numbers in its memory and the software treats them
appropriately. As a result, computer users can employ the full
range of alphanumeric text without special effort.

Since computers can store words, they can store their definitions as well. With programmed access to those stored definitions, some computers can keep up one end of a simple conversation. Unlike the definitions found in the dictionaries that
human beings use, however, a computer's definitions can't all
rely on words alone. Suppose that computer needed the meaning of "high" to answer the question "How high is the Empire
State Building?" If the definition for "high" was "tall," for
example, it would next have to look that word up, and so on
until it got to a word whose definition was "high." No reason
to stop there. The system would go back to the list endlessly—
looking up the definitions of words needed for earlier definitions.

When I came to America as a child and started school, I was
surprised to find a dictionary *all in one language.* The only
dictionary I had ever seen provided definitions of unfamiliar
(English) words in terms of (German) words that I could understand. I couldn't imagine what good English definitions of
English words could be to anyone. After a while, I got used
to the idea that an all-English dictionary could help someone
who had access to the meaning of at least some English words

from experience—like remembering the taste of an ice cream cone and someone saying "delicious," or connecting words like "bed" and "car" to mental pictures of familiar objects.

From childhood on, each human being develops a personal vocabulary based on the integrated sums of categorized experiences. As a child, only my German vocabulary had an experiential foundation. I started learning English by memorizing the English equivalents of familiar German words. At first, I could only express myself by constructing a German sentence in my head and translating it word by word. In the reverse direction, I translated the English words I could recognize into their German equivalents in order to understand what someone else was saying. Within a few months, however, I grew to use English directly, thanks to the miraculous language-learning ability young children possess.

Learning French in college took much more effort. Again I began with word-by-word translations. While I still have to rely on this awkward method for a large part of my French vocabulary, traveling in France has helped me build a modest store of experience-based definitions in this language as well. The store contains some unusual words, like *dessaler* ("to soak"), learned from overturning a sailboat in the chilly waters of Lac d'Annecy and having to ask a French clerk for my room key with water squishing from every piece of clothing. Most of the French words I don't first translate into English come via the more mundane route of frequent practice, such as the digits on a telephone dial. These direct contacts with a French environment allowed me to connect personal experiences with a vocabulary in that language along with the two I had acquired as a child.

Just as human language understanding depends on access to experience-based knowledge, computer-based systems require their own equivalent of the stored experiences human beings use. Since people acquire nonverbal knowledge through physical encounters with their environment, disembodied computers confront system designers with a fundamental problem. The human brain gets experiences from its integration with the human body. Can programmers provide computers with adequate substitutes for these sensory links to reality?

In this regard, computers that manipulate physical objects may enjoy a potential advantage over the ones that deal with abstract information alone. When computers perform physical tasks, their actuators and sensors give them rudimentary equivalents of human manipulation and sensing capabilities. Robotics researchers at MIT, General Motors, Bell Labs, and elsewhere have had some success in programming such machines to "learn" about their environment through exploration.

At Bell Labs, for example, a robotics research group built a robot that can follow spoken instructions. The system consists of a computer, a speech-recognition module that converts spoken words into computer-readable text, several video cameras, and a mechanical arm. If ordered to "take the red cone and place it on the blue box," for instance, the robot will do so—provided those objects have been identified during the training period that precedes each work session.

The human trainer begins by placing a number of objects within reach of the robot's arm, walks to a control panel located at a safe distance, and starts the machine. The computer uses the video camera to locate each of the objects it does not recognize from prior training, and then moves the arm near one of them. It waits until the trainer has spoken the name of that object into the control microphone. Then it stores that information and repeats the process until it has a name for each object. The computer controls the degree of bending at each arm joint in order to place the business end of the arm at the desired location, where a gripper functions like a crude human hand.

This gripper employs tactile sensors for the same reasons that human hands need sensitive fingertips.

Given enough time and appropriate programming, the computer could probe the surface of every object within its arm's reach. The results of such measurements could allow the system to supplement the trainer's input with direct information about these objects. For example, the system could distinguish "hard" from "soft" objects on the basis of the amount of deformation caused by a given amount of pressure on their surfaces, or "heavy" from "light" objects by trying to move each one.

The ability to infer the properties of encountered objects ought to make such systems valuable for future robotic applications—such as sorting parts and assembling mechanical components—but it has not yet led to new ways of storing information in computers. Designers must still program the computer to organize the results of its explorations by tabulating the properties of the objects in question, thereby creating so-called *property lists.*

Computers (and their robots) work in an environment, such as the one created in our robotics research lab. While our computer could respond correctly to instructions on the basis of its "firsthand experience," the list of properties stored in its memory might just as well have been composed by the programmer and typed into the computer directly.

Since the earliest days of programming, property lists have provided a convenient means for representing information about people and things. Our robotic computer would represent its self-generated information in that same form because computer memories have been specifically designed to store strings of alphanumeric characters and the connections between them (such as the information needed to organize a group of items in the form of a table).

For many applications, therefore, property lists represent the closest thing to experience-based definitions that today's computers possess. The recorded connections between the entries stored in a computer's memory give the system a rudimentary capability for representing nonverbal knowledge. For example, imagine that our robotics lab contained a table and on it a block of wood. Now, consider the question: "What is the height of the block at the center of the table?" A computer system might deal with this question if its property list for the object in question contained an entry for height. That entry represents the computer's connection with physical reality for this particular attribute—whether the system obtained the information by means of its own sensors or received it as a typed string of characters. These lists and the links between them constitute a powerful knowledge representation technology, forming the basis for most of the language-based transactions between computers and their human users.

As with most other kinds of technology, nature's own knowledge representation makes our technological contrivances look puny by comparison. Nature even manages to represent subtle concepts in a single molecule of DNA. The DNA molecule embodies the genetic blueprint from which we derive all the inherited characteristics we bring into the world at birth. This genetic code is "written" in the sequential placement of the four kinds of nucleotides stacked in the DNA molecule, forming a long twisted ladder with four different kinds of rungs.

The DNA blueprint must encapsulate not only every physical attribute of our bodies, but also the concepts that underlie our inborn instincts. For example, nature has managed to capture the notion of "height" in mammalian DNA.

Psychological tests show that human beings, and most other mammals, fear heights at birth. A newborn mountain goat placed in a glass-floored experimental chamber set on a surface with a large hole in it will carefully avoid the "dangerous" parts of the room even though it has no firsthand knowledge of falling. Somehow nature reduced the concept of height to a particular sequence of rungs on its DNA ladder. The message contained in this sequence tells living creatures to keep away from cliffs, regardless of what language (if any) they may learn later.

How does nature manage to "teach" a single molecule about height? This particular molecule forms the template for an integrated biological system, a living being. The knowledge contained in the template transfers to the living being as part of the physical design. Building a human being includes connections between the senses and the brain. This "wiring" provides a means of creating reflexes.

In this case, a combination of reflexes can implant the following behavioral algorithm (or sequence of operations): "If eyes focus on a field of view with the pupils apart while looking down, then *watch out!*" Such an algorithm builds the abstract concept of height into the body itself. Later, when the owner of the body learns a language, a word like "height" connects to that sensation.

The experience-based definitions that underlie human language understanding differ dramatically from our computer-based technology. Human intelligence moves outward from the physical encounters of the body with its environment, and is able to create abstract knowledge autonomously by integrating experiences in context. On the other hand, a computer must depend on human intelligence to categorize and connect the discrete pieces of information stored in its memory.

For example, the attribute of "height" can refer to distance between the top of a building and the sidewalk, or the elevation of its immediate surroundings—as in "How high is the Empire State Building?" and "How high is Bernie's ski lodge?" respectively. The designer of a computer system capable of responding to these questions—and only to these questions—must provide two "height" entries, along with explicit rules for deciding which one matches the context of each question.

It ought to be clear now that programmed machines can answer questions only within the context established by their programmers. Some users tend to overlook the limitations behind a computer's answers and give those responses an undeserved credibility. Avoiding such pitfalls calls for keeping the differences between human and machine-based vocabularies well in mind, especially whenever we enter into "conversations" with our machines through the exchange of verbal symbols.

Within the limits of a particular context, the computer's ability to handle language offers its human users much valuable help in dealing with the various kinds of textual information encountered in daily life. But that's not the whole story. Most of the information human beings use comes to us as visual images. To become truly useful helpers, therefore, computers must supplement their numerical and textual prowess with the ability to handle pictorial information as well. That's easier said than done, as is illustrated below.

Images

So much of our information flows to us through our eyes that we often say "I see" by way of meaning "I understand." No doubt, our powerful ability to extract data from visual images underlies the pervasive use of graphics as a means of conveying information in our society. For example, imagine having to use words alone to describe the information embodied in the wiggly line that traces past behavior of the Dow Jones index.

Computers can help our visual information-processing skills in two ways. First, by improving human access to graphical information, and second, by reducing the need for human labor in routine image-recognition tasks. While computers can contribute in both these areas without processing the images themselves—as in automating the storage and distribution of microfilmed records—most image-related tasks require image processing. And that's a tough job.

To give the computer the ability to process an image, the system designer must first devise a scheme for describing, or *representing,* that image in a format suited to the computer's memory. Using a "pencil-and-paper" analogy for the computer, we can regard this memory as a large ledger with spaces for tabulating numbers or alphanumeric characters. As with words, this representation task consists of converting the image into a numerical form, a form that the computer user can process by writing programs to operate on the numbers stored in the memory.

Imagine creating simple figures—such as this lowercase "e"—by rearranging a checkerboard made of movable black and white tiles. If we assign an address to each square—from

1,1 for the upper left to 8,8 for the lower right, for example—and record the color of the tile at each address, we can represent each of these figures of 64 squares in a tabular form suited to a computer's memory. We can extend this scheme to depict more fine-grained figures by using a larger number of smaller squares, up to many millions. Modern computer graphics systems normally devote as many as one million "squares"—called

picture elements, or *pixels* for short—to each picture in order
to produce high-quality images.

In addition to providing a medium for the creation of hand-
crafted images, computers can also acquire images from the
outside world, usually by means of an appropriately interfaced
video camera. In this system, an electronic device carries out
a sequential scan of the image. It measures the light intensity
at each of a preselected number of locations, thereby producing
a grid of picture elements like the "tiles" described above. The
computer stores this "digitized" image by recording the light
intensity of each pixel as numbers in its memory. Reversing the
process—reading the stored numbers and using their values to
modulate the brightness of a grid of dots on the screen of a
cathode ray tube—reproduces the *digitized images* on the computer's display.

Recent advances in office machine technology have greatly
increased the utility of digitized images. They now aid productivity by providing a versatile alternative to paper documents.
Most attacks on the high cost of paper files fail to create "paperless offices" because traditional computers themselves impose
rigid formats on their users. These text-based systems require
their users to type all documents on approved terminals, because they can store only the typed characters themselves.
Hand-written notes, figures, marginal annotations, signatures,
and mail from outside the organization generally call for awkward special treatment. In contrast, a system based on digitized
images can accommodate virtually any paper document, regardless of format, and still offer the advantages of electronic
storage because it stores an exact image of the original document itself.

Electronically stored documents never stray out of reach
because a copy of the "original" remains available at all times
to all potential users. To make effective use of this emerging

capability, however, users need access to high-resolution displays and printers to provide acceptably legible images, as well as high-speed data links to the storage facility (because of the enormous amount of data involved). Until recently, the cost of these items discouraged their use. Now, however, prices are dropping for electronics, laser printers, and digital communication systems. The cost barrier will soon be a thing of the past.

As digitized image storage and retrieval systems get cheaper and more flexible, I am personally looking forward to integrating loosely formatted information—like scribbled notes and sketches—into the computer-based office system we now use. An important part of this integration will depend on the computer's ability to extract meaning from the images we hope to store in the files under its control. For example, consider the case of the manuscript of this book. It is currently filed in the conventional manner as a set of characters typed into our office computer. That computer can provide me with all sorts of help, such as key-word searches, spelling checks, word processing, and even some help with sentence structure. On the other hand, if I had stored an image of each typed page instead, the computer couldn't operate on the contents unless it recognized the alphanumeric characters those images represented. Unless the computer could read that text, some human being would have to sit down with a printed copy and type it back into the word processor in order to modify it.

How would a computer "read" text from a digitized image? Some recognition programs compare the group of pixels representing each character against a set of stored examples—such as a more finely defined version of our checkerboard "e." After aligning each pair and scaling them to an equal size, such a system would measure the degree of agreement by counting the number of matched and mismatched pixels. The stored examples act as templates. Consider the following letters, however.

$$e \quad e \quad o \quad c$$

Our left-hand "e" has more pixels in common with the "o" and "c" than it does with the italicized "*e.*" Thus, the character recognition method I just described works only with a fixed font of letters. That is why many recognition systems don't rely on template matching. They use more general descriptions instead.

Consider: "A horizontal bar is connected at both ends to a downward-facing arc. The left-hand end of this same bar is also connected to the left-hand end of an upward-facing arc whose other end points toward, but does not touch, the right-hand end of the bar." Unlike the template method, such a description (recast in mathematical terms) would identify the similarity between "e," "*e,*" and "e," even though it would miss many of the other "e"s that people encounter every day.

Since even the various images of a given character from the same font rarely match each other exactly, ambiguities often arise. If, for example, the typing process omits the horizontal bar from a lowercase "e," the identification system could produce three possible matches, the intended "e," an "o" with the lower right corner missing, or a "c" with a long upper arc. In such cases the system might resort to a dictionary, using the other characters in the word to resolve the ambiguity. Suppose the "e" sat between a "b" and a "t." The system would find "bet" but not "bot" or "bct" in the dictionary and therefore reject the last two possibilities. In the case of "doe," "doc," and "doo," a resourceful program designer could extend the system to seek matches in a broader linguistic context.

Outside the office environment, the capability of matching digitized images against a stored set has found useful employment in dealing with physical objects, as in the Bell Labs robot I described earlier. Unlike the relatively straightforward bookkeeping that underlies template matching, extracting a three-dimensional object from a video picture calls for a great deal of sophisticated effort—all the more frustrating to computer scientists because biology does it with such apparent ease.

Recognition programs usually begin by scanning the image to find *edges*—well-defined boundaries between areas of different colors, brightnesses, or textures. These edges provide a basis for creating a three-dimensional "stick figure" through application of the rules of perspective. The computer stores these stick figures, sometimes supplementing them with information about surface color and texture in order to provide a basis for recognizing the same object in the future. Aligning each new stick figure against similar images from the stored set would enable such a system to locate and compare the key features of the scanned object. In that way, for example, the robotic computer I described earlier could recognize an object it had previously encountered from a different aspect.

While such systems can give human beings relief from tediously repetitive, routine image-manipulation tasks, they currently offer little of the flexibility associated with human pattern-recognition skills. Most systems would interpret an unexpected shadow as a different surface, for example. The present capabilities of machine-based vision limit its use to situations in which the system can refer to a prearranged set of models of the objects it must recognize. Even then, reliable operation demands careful illumination and extensive pretesting.

As for the more difficult problem of interpreting images in general, the computerized extraction of self-evident (to human eyes) features from an everyday photograph remains a topic for Ph.D. theses at our leading universities. Decades of progress in vision research have given us far more insight into *what* the human brain does in analyzing images than into *how* it manages to do it.

Conclusions

To sum up, I think it's fair to say that, apart from a handful of specialized pattern-recognition tasks—such as identifying a fixed font of characters in a document, or selecting objects from a factory conveyor belt—the extraction of useful information from pictures remains beyond today's computing methods. Furthermore, without some unexpected breakthrough, I see little chance of dramatic improvement in the near term.

While we humans routinely identify the various objects that appear in our fields of view with unconscious ease, we have only the tiniest understanding of how our eyes and brain accomplish this task. As a result, we learn visual recognition by studying examples rather than by memorizing rules.

James Audubon taught others to recognize birds by painting examples, expecting the reader's common sense to fill in the information he couldn't verbalize. My Ph.D. thesis adviser, Charles Townes, is a dedicated birder. He has traveled as far as the Himalayas in pursuit of his hobby—sighting over two thousand different species over the past fifty years. Few of these birds posed exactly as the guidebook showed them. Even though each of these sightings occurred under an unpredictable variety of conditions, he was able to connect their glimpsed images with illustrations made under totally different circumstances.

Consider the difference between writing a program to recognize the letter "e" and one to identify a pileated woodpecker. While I expect the former to be flat on the page in an upright position, the latter can assume a number of different poses, sitting, flying, or preening feathers—at unpredictable angles, at different distances, and under different lighting conditions. What happens when a branch gets in the way? I can't even imagine trying to produce a complete set of reference images for each conceivable encounter, let alone provide explicit instructions for making the needed comparison under each possible sighting circumstance.

Since computers need this level of explicit direction to accomplish an image-recognition task, they mainly bring images to human attention rather than extracting meaning from them. I compare it to having the foreign language section of a library presided over by someone who understands only English. Such a librarian could help researchers in only a limited way by relying entirely on a book's spine, or its call numbers, rather than on any direct understanding of its content.

On the other hand, I am sure that this situation will change drastically when (and if) we can devise a computing architecture that learns from examples, the way humans do, rather than merely obeying instructions. Nature produces countless billions of such "computers" in the brains of living creatures. Unfortunately, we can't yet match that capability with present-day programming methods. Until some better idea comes along we can only expect "clerical" help from computers in handling image-based information.*

When we move from pictures to numbers, however, we encounter quite a different situation. Machines rarely have trouble extracting *all* the information numbers contain. But, as my "transmission parts, yes; boutique gifts, no" story illustrated, numbers alone can't say anything about the real-world contexts in which their users seek to employ them.

Of the three types of common symbols, therefore, I think it's fair to say that only words offer us a means of describing a wide

---

*Such help can be extraordinarily valuable, as in the computerized storage and retrieval of massive document files. As we explore the computer's ability to match human information-processing capabilities, we shouldn't overlook the fact that these machines have much to offer society without emulating humans.

variety of information in a format that machines can interpret. As a result, I will be focusing on the mechanization of language understanding as the central theme in much of what follows.

The power of human language to teach, inform, and persuade has made a high level of literacy an important goal throughout the world. We need a similar level of literacy in our computers. Because today's computers don't understand human languages, today's computer users must make their wishes known to a human programmer for conversion into an explicit series of mechanistic instructions—or settle for whatever ready-made software product comes closest to meeting their needs.

Someday, users might skip that step, and send their requirements directly to a machine in everyday language. In that case computers would extract enough meaning from a user's words to create the corresponding machine instructions. That extraction of meaning calls for grammar, logic, and the other methods humans use to accomplish this task.

Chapter 3

Rules

*Plura faciunt homines e consuetudine, quam e ratione.**
—GEORGE MACDONALD FRASER, *Flashman and the Redskins*

RULES UNDERLIE all languages, the *special-purpose languages* (like Basic and Fortran) that we create for computing no less than the *natural languages* (like English and Japanese) that we use in everyday life. But there's a big difference. In computing languages, the rules come first. A designer must feed a computer an explicit set of carefully crafted rules for extracting meaning from every possible sentence that each machine will handle. Human languages, on the other hand, have grown up organically. People constantly create new word structures to meet their spur-of-the-moment needs in dealing with the unpredictable course of everyday life.

The "use now, analyze later" methods employed in the creation of natural language lead to an imprecise and incomplete rule structure. Since grammarians—not to mention ordinary mortals—must infer their rules from existing usage,† they can never hope to cover every combination of words that makes sense to people. In addi-

*"Men do more from habit than from reason."
†In some cases government-supported academicians try to legislate word usage, such as the French *le logiciel* (the official equivalent of the more frequently used *le software*).

tion, ingenious verbalists and writers frequently contrive successful evasions of existing rules—or lighthearted strict adherence, such as Winston Churchill's "Ending a sentence in a preposition is a practice up with which I will not put!"

This lack of precise rules allows natural languages to grow and to adapt flexibly to changing situations, but it also makes computer-based analysis a difficult undertaking. Consider the sentence: "The total is five times thirty plus thirty-five." The total could be $1.85 (as it would be if I were buying a half-dozen bagels, one of which cost five cents extra) or $3.25 (for five bagels, each with a 35¢ portion of cream cheese), depending on whether one adds before or after multiplying. Computer languages avoid this ambiguity by setting hard and fast rules about such matters. In most such languages, computers multiply before they add, unless told otherwise by parentheses, as in $A = B \times (C + D)$. No computer language would leave the interpretation to chance. Otherwise, a program would need to halt and wait for clarification every time it encountered such a statement.

In contrast, even the most pedantic grammarian wouldn't try to impose a "multiply-before-add" rule upon all the users of the English language. People manage on their own without such "help." When necessary, they use intonation, stress, or timing to disambiguate their speech—as in "five times [pause] thirty plus thirty-five." That's perfectly satisfactory for human conversation. But what's the poor computer to do when confronted with a text created by and for users who enjoy all the benefits of "common sense"? Clearly, computer-based natural language systems require a guide to the methods people use to extract meaning, a guide that is as explicit as the scientific understanding of language can possibly make it.

Linguistics

Natural languages serve as the primary means of expressing knowledge in almost every facet of human endeavor. Even when other forms of expression predominate—as in accounting, art, or music—language acts to make specialized material more generally accessible and expressible, playing a role like that of the chairman's letter to stockholders in an annual report, or the program notes offered to concert audiences. Both act as a kind of translation. As a result, most computer-based information-processing tasks involve the use of language—from the initial statement of the problem to the generation of the final report.

Once computers become better able to extract information from natural language, many users will be able to skip over the arduous work of converting everyday phrases into the precise commands that today's machines require.* But most of that capability is still in the distant future.

Since human beings created natural language for their own use, its structure reflects the integrated and multilevel way a human brain processes information, rather than the neatly sequential messages that machines usually generate. This disparity has given scholars much to chew on for several decades, and some—like MIT's Noam Chomsky—have made prodigious strides in laying the groundwork for linguistic analysis. In what follows, I will trace somewhat simplified versions of illustrative examples that Chomsky uses. They show how the information structure of human language creates formidable difficulties for designers of computer-based language-understanding systems.

*I don't expect the need for professional programmers to disappear. Programming in computer language will presumably yield performance advantages—much as racing car drivers employ stick shifts rather than automatic transmissions.

Language analysis must extract meaning at three levels: the *semantic,* which relates words to things, giving them meaning; the *syntactic,* which relates words to other words, expressing connected meanings such as the action of one thing upon another; and the *pragmatic,* which relates these words to the context that encompassed their creation.

My friend Mitch Marcus, the Bell Labs linguist who first introduced me to this subject, began my lesson with the sentence "The boys refused the tea and coffee." It wasn't hard for me to believe that semantic analysis could attach meaning to each word and pass that information on to some kind of syntactic analyzer, which would then diagram the sentence. ("The sentence begins with an article which modifies the noun immediately following it. That noun acts as the subject of the verb whose object is a pair of nouns modified by an article and connected by a conjunction.")

The straightforward nature of this sentence lends itself to analysis that extracts meaning in a readily explainable (and easily programmed) series of steps. That was the good news. But, as Mitch explained, people didn't design natural language to suit formal analysis. Instead they reuse the same structures in as many different ways as suit their intuitive ability to tell these variations apart.

To illustrate his point, he put the word *have* in front of our sentence, making it a question: "Have the boys refused the tea and coffee?" Next, he changed the tense of the verb *refused* to give us a command: "Have the boys refuse the tea and coffee!" Note that the meaning of the first word depends on the fourth, so the semantic analysis program would have to look ahead to the verb before assigning a meaning to *have.* Furthermore, the speaker might wish to identify the subject more specifically: "Have the boys who have to go to sleep now (because they have to handle early morning duty at the campfire) refuse the tea and coffee." In this case, finding the right verb requires identification of the intervening dependent clause (or clauses), so the semantic analyzer would need help from syntax in order to accomplish its job.

Finally, Mitch had me consider the sentence pair: "The boys refused the tea and coffee because they were too warm" and "The boys refused the tea and coffee because they were too cold." In the first sentence, the clause concerns "the boys," who were too warm to want (normally) hot beverages. In the second sentence, however, that same clause concerns "the tea and coffee," which had cooled below an acceptable serving temperature. Consequently, extracting the meaning represented by these words calls for a pragmatic understanding of what people drink—an understanding of culture.

In principle, each potential connection between the meaning of one word and that of another obligates the system to explore a large number of additional combinations. Since systems designers try to anticipate as many of these combinations as possible, "simple" definitions often become dissertations. Under "hot tea," for example, we would need to know not only the normal serving temperatures and the circumstances under which it is offered and accepted for consumption, but also references to iced tea as well. Furthermore, "hot" material might be "highly desirable," "stolen," or even "radioactive." No wonder people sometimes misinterpret what they read or hear.

While computer-based language-understanding systems make many more mistakes deciphering complicated sentences than people do, they do provide useful help in restricted applications—such as the natural language interface to a computer-based do-it-yourself airline reservation system. The limited vocabulary and syntax required in such applications allows the system to offer more robust service than it could achieve with the use of unrestricted language. Despite their limits, such systems can often help their users in unexpected ways.

"Which convertibles have repainted roofs?" A skeptical senior manager of a major automobile manufacturer had asked that question during a demonstration of a computerized natural language query system. The system was supposed to provide flexible access to the voluminous sales and service records of the prospective customer. While it wasn't the sort of query the engineer had hoped for, he had no choice but to seek an answer from the language query system. He typed in the five words, hit the return key, and hoped for the best.

Some moments later, the answer came back, a neatly typed list of several vehicle serial numbers. The questioner and his colleagues chuckled at the computer's "silly" response. But the engineer didn't give up easily. He asked his system a few more questions and soon regained the audience's undivided attention. All the listed convertibles had had their roofs "repainted" by the same dealer. Furthermore, that same dealership also led its region in the number of sedans, hardtops, and station wagons that "required" repainting—all at the manufacturer's expense. By that time, the system had sold itself.

Because a skeptical questioner had stumbled on the right question, pertinent information emerged from the mass of stored data. But random coincidence can't offer much of a basis for extracting needed information from computers. We would like to exploit the flexible generality of natural language to describe a broad topic of interest ("Which dealers consistently bill us for above-average expenses?") and to generate the precise instructions that govern mechanistic behavior ("List all dealers in order of their touch-up charges."). Clearly, some person—or some formal mechanism—must fill in the missing steps through the application of reason. And that brings us to my next topic.

Logic

Like many other fields of scholarship, the systematic analysis of language owes much of its foundation to ancient Greece. Most notably, Aristotle first described language in terms of subject and predicate—as well as the parts of speech—over 2,300 years ago. Early Greek grammarians understood the trade-off between flexibility and accuracy. They saw how the imprecision of natural language undermined its utility in mechanistic applications such as formal reasoning. Their insights led to the ultra-precise "language" we call *logic.*

While humans need not confine themselves to formal logic when solving problems, today's computers have no viable alternative. Programmers must provide machines with either a sequence of step-by-step "do this, then do that" instructions or a set of unambiguous rules to govern their behavior. Logic provides the fabric for dealing with such rules.

As the father of logic, Aristotle circumvented the imprecision of natural language by creating more precise ways of reasoning. For this alone, Aristotle would have achieved immortality. The methods of *Aristotelian logic* provide the foundation for all of our modern world's computing languages. As a result, computer-based decision-making reflects both the strengths and the drawbacks of logical systems.

Aristotle and his fellow logicians confined themselves to a small number of unambiguous constructs, such as, "If A, then B": the truth of one fact ("A") implies the truth of another ("B"). This celebrated rule gives Aristotelian reasoning the power to establish facts through inference. Reasoning provides

a method for adding value to information by increasing the
number of interesting facts (facts that affect decisions) at one's
disposal.

For example, imagine yourself preparing to cross the street.
Reasoning could permit you to infer "The street is safe to cross"
from "The light is green for pedestrians" by the use of a sequence of connective rules, such as:

>If the light is green for pedestrians,
>>then the light is red for vehicles.
>If the light is red for vehicles,
>>then the vehicles stop.
>If the vehicles stop,
>>then it is safe to cross the street.*

In this way, logic provides an explicit way of following the path
from observation to conclusion.

With its elegant simplicity, logical reasoning appears to provide a foolproof mechanism for making decisions. That's why
most people regard a "logical mind" as a positive attribute. But
there's more to the story. How often do real-world problems
bring about the cut-and-dried reasoning I set up in the traffic
light example?

Clearly, you must first come to the conclusion that "the street
is safe for pedestrians" before attempting a crossing. But when
is the last time you held a mental dialogue with yourself under
such circumstances?

On the other hand, let me present you with a puzzle that calls
for much mental dialogue. Find the numerals represented by
the letters in the following addition puzzle:

*Crossing a real street involves a host of factors and would require a far larger
set of rules. Some vehicles don't stop for lights. The street, or the opposite
sidewalk, might present hazards—potholes, muggers, poison ivy, or a curb
too high for the baby carriage you happen to be pushing.

$$\begin{array}{r} \text{HOCUS} \\ +\text{POCUS} \\ \hline \text{PRESTO} \end{array}$$

The instant you begin to work on it, you can *hear* yourself think. In this case, there appears to be no alternative to the conscious use of logic in order to test and prune the various possibilities before arriving at the final answer.

The fact that "PRESTO" has one more digit than "HOCUS" or "POCUS" provides an important clue. If the addition of two numbers results in a sum with an extra digit, then the leading digit of that sum must be a 1—since the sum of any two digits plus a carry can't exceed 19. If P is 1, then H must be 8 or 9, and R must be 0. If H were 8, O would have to be greater than 4 (in order to produce a carry). Furthermore, O must be even in order to be the sum of S plus S. Therefore O must be 6 (the only remaining even digit greater than 4).

H=8, O=6	H=9, O=4	H=9, O=2
86CUS	94CUS	92CUS
+16CUS	+14CUS	+12CUS
10EST6	10EST4	10EST2

For H to equal 8, therefore, S must be 3 (since 8 is already taken). But if S were 3, C would have to be either 1 or 6, both of which are also taken. That leaves us with H equals 9. By this time, anyone but a combinatorial genius will probably have lost interest in the argument or resorted to pencil and paper.*

*In writing this section, I remembered the problem but not the solution. So I had to work it out again for myself. My first attempt took two sheets of paper, and most of a shuttle flight between Newark and Washington, D.C. My son found a flaw in that argument, so I spent another hour—with some help from a friend—getting it right. Here's the rest of my analysis.

If O is 4 then S is 2 or 7. If S is 2 then C must be 6 (since 1 is taken). But that gives us a carry, which makes E equal to 9. Since 9 is unavailable, we

Compared to a computer, my problem-solving performance was ridiculously slow. A computer program that merely tried every digit in ascending order, starting in the upper right-hand corner, would finish the job in a few seconds. I took over an hour. Furthermore, I would have been even slower if I hadn't cut down my search by identifying the 1 at the start.*

Given this level of performance in a problem which had only about twenty intermediate steps, how does anyone have time to analyze all the factors involved in crossing a traffic intersection? Are we better at logic than the above example suggests? Modern psychology offers a different explanation.

Dr. P. C. Wason, a British psychologist, suggests that most people *overestimate* their own ability to apply logic to problem

---

must next try S equals 7, which leads to C equals 3 (since C equals 8 would also produce a carry), making E equal 8.

$O=4, S=7, E=8$	$O=2, S=6, C=3$	$O=2, S=6, C=8$
943U7	923U6	938U6
+143U7	+123U6	+128U6
1087T4	1046T2	1056T2

But for $U+U=T$ to produce a carry, U must be either 5 or 6 (the only remaining numbers greater than 4), but these values would make T 1 or 3, respectively, both of which are already taken.

That brings us to O equals 2. Since 1 is already taken, S must be 6, which makes C either 3 or 8. In both cases (as shown above), U must be less than 5, in order not to produce a carry. But all the numbers less than 5 have already been used in the former case. In the latter case, however, both 3 and 4 are still available—making T 7 or 9, respectively. Since 9 is already taken, only 7 fits.

As a result, "92836 plus 12836 equals 105672" is the only possible answer.

*People don't normally write numbers with zero as the first digit. When I later considered P to equal zero I found two additional answers: "86173 plus 06173 equals 092346" and "46193 plus 06193 equals 052386."

solving.* To support this assertion, he offers a series of logically structured problems that illustrate the kind of unexpected difficulties many people experience in applying the theorem-proving methods of formal logic. The following pair of problems exemplifies his ideas. Read the following paragraphs carefully and see how well you do.

The first problem concerns four pieces of cardboard with numerals printed on all eight faces. Imagine these four cards lying on a table top in front of you with the visible faces displaying the numerals 2, 4, 3, and 7, respectively. If you were asked to verify the truth of the proposition "If a 2 is printed on one side, then a 7 is printed on the other" for these four cards, which ones would you have to turn over? (If you turn any over unnecessarily, you lose, so you may wish to review your analysis once more before reading further.)

The face behind the 2 should have a 7 on it; that's what the rule says. Accordingly you must turn over the 2 to make sure that a 7 lies behind it.

If the face behind either the 4 or the 3 had a 2 printed on it, the rule would be violated, because a 2 on that side would call for a 7 in place of the 4 or the 3. You must therefore turn over both the 4 and the 3, to make sure neither of their hidden faces has a 2 on it.

Few of the people who take this test have trouble with these first three cards. When it comes to the fourth card, however, a large majority—typically five out of six—mistakenly conclude that they must turn over the 7 to make sure there is a 2 on the

*P. C. Wason and P. N. Johnson-Laird, *The Psychology of Reason: Structure and Content,* Harvard University Press, Cambridge, Mass., and London, Batsford, 1972.

other side. They fail to realize that there is no way of violating
the rule if there is a 7 on one side of the card. (The following
example explains why.)

People have much less trouble with this next problem. Again
we have four cards, each with one face showing, but this time
they are pinned to the jackets of four people leaving a bar. One
face displays the bearer's age, and the other presents a truthful
record of that person's recent beverage consumption. The visible side of the first card labels the first person a "beer drinker,"
while the other three cards list the ages of their wearers as 18,
17, and 23, respectively. Which cards would you have to turn
over to test the rule "If someone is a beer drinker, then that
person is over 21"?

Testing the validity of this rule requires getting the beer
drinker's age, which means turning over the first card. Next, the
two under-21 customers could have violated the rule by drinking beer, so each of their cards must be checked for beer drinking, and turned over also.

This leaves the 23-year-old. Few people—less than one out
of seven in fact—make the mistake of concluding that testing
the rule requires a check of the habits of someone who is over
21. Yet "over 21" occupies exactly the same position in the
problem as "7" did in the earlier one.

> If a 2 is printed on one side,
>   then a 7 is printed on the other.
>
> If someone is a beer drinker
>   then that person is over 21.

Most people find it hard to understand that any card with a
7 on one side satisfies the second rule, whether or not it has a
2 on the other. On the other hand, few people have trouble

understanding that someone over 21 satisfies the second rule, whether or not that individual drinks beer.

The dramatic correlation between content and success rate shown in this pair of "If A, then B" examples wouldn't exist if people had made correct use of logic to solve such problems. The "logical" approach calls for use of the rule, with appropriate substitutions for A and B in each case. If the people tested had actually employed the rule to solve these problems, they would have done equally well (or equally poorly) in both cases, since the rule's operation doesn't depend on what one substitutes for A and B.

Psychologists cite the dramatic difference in the two success rates (less than 20 percent correct with the numbered cards, versus almost 90 percent with the drinking-age cards) to demonstrate that people don't normally use logic to solve problems. Most people, it seems, haven't learned to make use of logical methods. They depend instead on prior experience, succeeding in the cases where they have it (like drinking age), and struggling where they don't. Nor does profession seem to matter. Most scientists and engineers do no better with this problem pair than my nontechnical friends. (Only professional mathematicians demonstrate enough facility with logic to avoid the trap consistently.*)

Society rewards, and evolution favors, successful problem solving. As a result of the cumulative effect of these forces, we can expect today's most successful problem solvers to have made use of the best methods available to them. Outside of a few specialized areas like mathematics, however, logic seems little used in attacking problems.

Does this apparent "aversion" to the general use of logic imply a shortcoming in human problem-solving power? I doubt

---

*They make a clear distinction between "*if*" and "*if and only if.*" Try substituting the latter expression in the examples.

it. After all, the fact that mathematicians make logic a full-time occupation (and appear to have fun doing so) demonstrates that the capability exists. But evolution and environment didn't act to emphasize it in the general population.

We must make some use of logic to deal with the rules that living in society imposes upon us, but we use it sparingly. Even in the areas such as law and accounting, in which logic provides the basic decision-making tool, successful institutions make provision to limit the reliance on logic alone. For example, the people who write laws try to build self-consistent and comprehensive rule structures to govern human behavior. Even so, our legal system employs legal precedents and judicial judgment to temper the literal analysis of "the letter of the law." Successful businesses do the same thing. Take Harold Geneen, who probably did more than any other business leader to establish the modern system of corporate financial accountability.* Geneen balanced his insistence on "hard numbers" with an equal emphasis on regular face-to-face meetings with the people involved.

As we undertake our exploration of the ways in which computers can aid decision-making, I need to show you why aggregated human experience so often chooses alternatives to the rule-driven mechanisms our machines employ. I will do that by taking a look at a pair of "detectives" in action: first, a computer-based expert system, followed by an old-time sleuth from the movies.

Deduction

In my earlier crossing-the-street example, logic led us in a straight line from "The light is green for pedestrians" to "The street is safe to cross." The real world, on the other hand,

*See, for example, R. J. Schoenberg, *Geneen,* Warner Books, New York, 1985.

usually forces us to choose between a number of alternatives, much as in a detective story. While few of us concern ourselves with the real-world counterparts of mystery-story detectives, we do encounter deduction in many everyday situations.

Let us suppose my car's engine refused to start some morning. Before calling for a tow truck, I would review possible causes to see if I could manage to fix the problem myself.

The starting process goes as follows: The act of turning the key closes a switch that sends current from the battery to the starter—an electric motor that turns the engine's crankshaft through a few revolutions. The rotation of the crankshaft, in turn, drives the fuel pump, which moves gasoline from the tank through a fuel filter to the carburetor (which mixes the gasoline with air for burning in the engine's cylinders). As the crankshaft rotates, it also operates the intake valves that allow the fuel-air mixture to enter the cylinders, as well as the exhaust valves that control the flow of burned fuel from the cylinders to the exhaust system. The crankshaft also drives the distributor which causes the ignition system to send a properly timed sequence of electrical pulses to each of the spark plugs in order to ignite the fuel in each cylinder at just the right moment.

With so much going on, I would need to start my search by organizing the possible causes of a malfunction. I could organize this information by creating a checklist of "suspects" in each of the major problem areas—starter system, ignition system, fuel, and mechanical parts—as follows.*

Starter system
  battery
  ignition switch
  starter motor

---

*This list, as well as the similar material that follows, is much abbreviated to save space.

Ignition system
  distributor
  wiring
  spark plugs
Fuel system
  carburetor
  fuel pump
  fuel line
Mechanical
  valves
  rings
  valve lifters

Clearly, finding the culprit calls for a better search strategy than merely running down such an arbitrarily ordered list and checking each item in turn. One handy way of setting priorities is to plot the hierarchy of possible situations in the form of a *decision tree,* as follows:

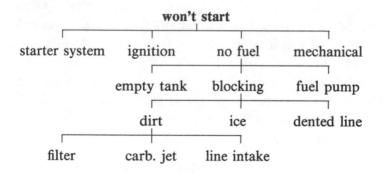

Starting at the top, each condition sits at a branch point that connects it to whatever condition might underlie it, layer by layer, down to the final causes (like a clogged fuel filter), which sit on the leaves at the tips of the "twigs."

With the tree as a road map, how would an expert mechanic

plot an economical path to the solution? Artificial-intelligence workers have addressed such problems by interviewing experts in the hope of mimicking their problem-solving abilities with computer-based systems. Because this programming methodology relies on interviews with human experts, its products are generally referred to as *expert systems.*

In addition to a set of problem-solving rules—obtained from the interviews with the relevant experts—each expert system requires a rule-control program, or *inference engine,* which selects the sequence of rules to be implemented and handles communication with the outside world. In an expert system on car repair, each of the branch points on our tree would be covered by a rule:

If the engine won't start
> then the engine isn't getting fuel,
> or the starter isn't working,
> or the ignition system isn't working,
> or the engine has a mechanical problem.

If the engine isn't getting fuel,
> then the line is blocked,
> or the tank is empty,
> or the fuel pump isn't pumping.

If the line is blocked,
> then there is dirt in the line,
> or there is ice in the line,
> or the fuel line is dented.

If there is dirt in the line,
> then the dirt is in the fuel filter,
> or in the carburetor jet,
> or in the line intake.

The sequential application or "firing" of these rules corresponds to a search path through our tree. In a real problem, the trunk splits successively into many more stems, branches, and twigs than I've shown. As the paths diverge time after time, the number of possibilities that need systematic checking mounts rapidly. To minimize the layout and cost of each search, expert systems employ mechanisms for setting priorities and taking advantage of possible shortcuts—for example, "If the day is unusually cold, begin by checking for ice in the fuel line."

If my cooked-up example had been part of a real expert system, a problem-solving session might proceed as follows. The user begins the session by selecting the symptom ("engine stopped") from a menu. The system matches "engine stopped" against the top line of the first rule and applies it. It finds that it has no stored information about the first cause, so it checks the rules and finds a match with the top line of the second rule, bringing it to "the line is blocked." Two more repeats of this process lead it to apply the fourth rule and "the dirt is in the fuel filter." This time the search fails to turn up either stored information or an applicable rule so it asks the user "Is there dirt in the fuel filter?"

A "yes" from the user solves the problem and ends the process. In a more typical case, a "no" would cause the system to scratch one possible cause and go on to explore the others. Typically, this system might hunt through hundreds of wrong answers before hitting on the right one. The biggest cost in carrying out an extended search comes from the tests the computer asks for along the way, rather than from the extra computations it performs. For example, suppose that the system had detoured to the ignition system before getting to the fuel line. If the computer had direct access to ignition data, the extra steps would have been almost invisible to the user. On the other hand, no one would enjoy being sent on a futile trip under the hood every time the computer needed to check out an extra possibility.

Computers rarely find shortcuts to problem solving as easily as people do. As a result, expert systems generally do better in problem areas in which extra tests don't cost much. In one such situation, the Digital Equipment Corporation uses an expert system called XCON (as in "expert configurator") to select the accessories (such as cabinets, power cables, and cooling fans) that each of its customized computing systems requires. Since the input itemizes all the computer components the customer wants, and the memory contains all the needed information on each component, the system can check alternatives (such as selecting an AC outlet box with enough sockets to accommodate the number of power cords that need to be plugged in) without troubling the user or incurring testing costs.

At the other end of the scale, medical diagnosis must operate in an environment of costly testing—in terms of both monetary expenditure and stress on the patient. Furthermore, medical tests often provide probabilistic rather than absolute answers. While Digital's XCON can stop applying rules concerning power once every plug has a socket, for example, a normal blood count diminishes but does not exclude the likelihood of internal bleeding (or preclude the need for further testing to "make sure"). Like human physicians, therefore, computerized medical diagnostic systems must deal in probabilities.*

To meet this need, computer scientists have extended their rule systems to handle probabilities. In medical diagnosis, two (much simplified examples of) such rules might read as follows:

If the patient has a serious stomach ulcer, then there will be an 80 percent probability of internal bleeding (i.e., in four out of five ulcer cases).

If the patient is bleeding internally, then there will be a 75

*Some data—such as the patient's blood type—aren't subject to random changes, unless the lab has made a mistake.

percent probability of a lowered blood count (i.e., in three out of four bleeding cases).

We can combine such rules just as we did the earlier true-false variety. In this case, however, they produce a probability instead of a "yes" or a "no."

If the patient has a serious stomach ulcer, then there will be a 60 percent probability of a lowered blood count (i.e., in three out of five cases).

Computer-based medical diagnostic systems employ rules like these to calculate the odds for and against the various diagnoses they present to their human users. Since further testing offers the possibility of reducing the uncertainty in these results, "automated" medical diagnosis systems (acting as advisers to physicians) generally tend to suggest every test their rules can access. This forces designers to limit the scope of such systems to carefully delineated problem areas. Finding these niches calls for human judgment.*

Just as a system designer's judgment selects application areas for computer-based logic, everyday human judgment routinely selects the portions of problems suited to logical attack by human intelligence. This need to guide logic with judgment provides a recurring theme in modern fiction, especially in detective stories.

Arthur Conan Doyle's Sherlock Holmes stories provide a convenient example of what I have in mind. Most of these tales unfolded according to a familiar three-part formula: Holmes encountered a seemingly inexplicable set of circumstances, figured out the underlying pattern, and then set a trap to validate his suspicions. That trap generally provided the story's

*That is, the ability to make successful decisions with incomplete information. While we have as yet little scientific understanding of the judgment mechanism itself, psychological testing indicates that humans seldom employ mathematical probability to make decisions.

climax. The guilty party usually walked into it and was turned over to the police.

The opening scene of each story presented Holmes with a loosely structured problem. In "The Red-headed League," for example, a shopkeeper bemoaned his loss of an unusual part-time job, which had just been mysteriously discontinued. During the interview, Holmes learned of a new clerk who had volunteered to work for the shopkeeper for half wages "so as to learn the business." Further investigation established the clerk's criminal record and the fact that the shop's cellar backed up against the rear of a bank vault on the next street. Mere coincidence, or the elements of a crime?

Establishing the suspect's guilt required proof, and proof calls for logic. Since the inconclusive nature of the evidence precluded the successful use of logic alone, Holmes first employed intuitive judgment to identify the subject (the suspected bank robbers) of a subsequent logical attack (the trap).

In setting a "trap," Holmes established an artificial situation whose outcome led to an unambiguous conclusion. In this case, Holmes and Watson (accompanied by a police officer and a bank official) merely sat in the darkened vault and waited for the anticipated break-in. Then, when Holmes and his companions caught the villains entering the bank vault through a tunnel that had been burrowed from the shop in the shopkeeper's absence, logic established guilt.

How does this story apply to computer-based decision-making? Like Holmes, we must expect judgment to precede logic regardless of what processors we use. Logic requires boundaries. We can only trust our rule-driven computers in areas that human judgment identifies as suited to logical attack. The Sherlock Holmes stories are believable because that's how the real world operates. Even the most astute human problem solvers defer the use of logic until the quality of the available information calls for it. Logic treats truth and falsehood presented as truth in exactly the same way. Computer people make the same point in the phrase "Garbage in, garbage out."

Properly used, logical methods play a vital role in human information processing. Logic provides mechanistic solutions to real-world problems—like diagnosing engine problems or scheduling an airliner. As we contemplate the mechanization of these methods, however, I must emphasize that no comparable machinery yet exists for judgment. Among other things, judgment must certify the validity of the hard facts that logic sorts out for us.

But there's a good-news side to this story, too. The strengths and weaknesses of rule-based problem solving lead naturally to a healthy division of labor between people and information technology. Since logic can't do it all, neither can logic-driven machinery. A computer can't handle the full-blown unrestricted problems that Sherlock Holmes faced any more than an automobile can find its own way to the supermarket. But that doesn't keep either one from being a useful helper in its respective tasks. Just as a realistic view of what automobiles ought to be called on to do is a prerequisite for safe driving, a similar understanding of information-processing machines will help us to use them properly as well.

Chapter 4

Machines

Tote that barge, lift that bale!
—RICHARD RODGERS and OSCAR HAMMERSTEIN

SOME MACHINES (such as tugboats and forklifts) move physical objects; others (such as clocks and computers) move symbols. Just as the first group extends the reach and power of human muscle by applying energy, the second group extends the reach and power of the human mind by providing needed information in a specified time frame, location, and form. To do so, these information machines must store, transmit, and process symbols. Information processing consists of converting one group of symbols into another in accordance with rules of logic and math, such as those I discussed in the previous chapter.

Most information machines are relative newcomers on the world scene. Throughout most of human history, machines performed physical tasks almost exclusively. As late as the mid-eighteenth century, for example, a single clock served the timekeeping needs of most small towns in Europe and colonial America. At that same time, however, most individual households had several machines dedicated to physical tasks, such as spinning wheels, grindstones, winches, and wagons.

By the second half of the twentieth century, the ad-

vent of modern telecommunications and the twin inventions of
the computer and transistor had changed that picture dramatically. Especially, advances in integrated circuit technology have
led to an explosive proliferation of microprocessors. Today,
these ubiquitous "computers on a chip" enjoy a plurality over
all other machines combined. In less than two centuries, the
role of information machinery has grown from the single town
clock to a position of dominance in a highly mechanized society. Whatever their primary task, few modern equivalents of
colonial spinning wheels, grindstones, winches, and wagons
function nowadays without one or more computer chips to
control them.

In addition to their ever-growing role in controlling physical
machinery, computers permeate virtually every information-
related task our society depends on. With computers and com-
puter-processed information shaping so much of work and everyday life, it helps to have some personal understanding of the
nature of information machines. In this spirit, I'll lead you
through the insides of such machines—past, present, and possibly future.

Yesterday

The mechanical manipulation of symbols began with clocks
and clockwork technology. Until the invention of the mechanical clock in the twelfth century, timekeeping depended on the
sweep of the earth's rotation—the motion of stars through the
heavens, or shadows across the face of a sundial. Replicating
those slow, steady motions with a reliable mechanical mechanism presented the key design difficulty, one that repeated attempts had failed to solve over many centuries.

I came upon the story of the solution to this clock regulation
problem in Vienna's Uhrmuseum, a small museum devoted
entirely to clocks and clockwork. Located in a beautiful old

building on a narrow side street in the oldest part of the city, the collection contains representative examples of clocks produced over the past eight hundred years as well as a few of their "ancestors"—including some examples of the ingenious rotation mechanism that inspired the first clockmakers to solve their long-standing regulation problem. That mechanism originated in, of all things, a machine for roasting meat.

The unknown inventor of this machine had found a way of turning meat slowly over a fire by rotating a spit at a slow but steady rate. (Just wrapping a light chain around the spit and attaching a weight to pull it would not have done the job. As the weight dropped, the chain would pull the spit around all right, but it would drop too fast. And the weight would be back on the floor almost as soon as someone rewound it.) In any case, this anonymous inventor managed to slow the rotation to the desired rate by devising a timing mechanism. A crude gear, mounted on the spit, engaged a small hook that caught each tooth as it went by. That catching action was timed by a swinging pendulum whose motion released the hook from one tooth and set it to grab the following one.

Imagine a hungry dinner guest watching the meat turn as the firelight cast shadows and reflected glints of light that moved slowly across the hearth. Shadows on a sundial? The stars in the heavens? Some image connected the passing of time with the motion of the spit, and that image gave birth to the idea for the mechanical clock (and led to the gear-based precursors of the modern computer).

The steady motion of the spit had to be slowed further to match the speeds needed for the hands depicting the minutes and hours adopted from sundial-based timekeeping. Proper clock operation demands an exact reduction of speed from one shaft to the next. For example, the shaft connected to the minute hand must rotate exactly twelve times as fast as the one on the hour hand. Suppose the minute hand rotated just 11.95 times faster than the hour hand. In that case, the minute hand would slip ten minutes behind the hour hand in less than two days. By noon of the second day the hour hand would point at the 12 all by itself, while the minute hand would point toward the 10, indicating ten minutes before the hour. Bringing these hands back into synchronism would mean turning the minute hand backward through two days' worth of revolutions, or taking the clock apart.

The high precision demanded by clockwork precluded the use of belts and pulley arrangements to keep the hands in synchronism. Because the amount of reduction depends directly on the relative diameters of the pulleys involved, the required accuracy imposes impossibly strict limits on permissible size variation. For example, the difference between reduction ratios of 12 and 11.95 equates to increasing the diameter of a broomstick by wrapping a single layer of human hair around it. Moreover, a practical clock would demand one thousand times better accuracy.

With gears, on the other hand, mechanical accuracies need merely assure that the teeth on each wheel mesh smoothly with those on the others. The operation of a geared clockwork depends directly on the *numerical* properties of its components (i.e., the number of teeth each gear has) but only indirectly on their *physical* properties (i.e., enough accuracy to keep the gears meshed). As a result, clock designers can expect hands to stay in synchronism until something breaks or wears out.

The single exception to the dependence on numerical properties involves the timekeeper that regulates the motion of the drive mechanism. Its period—or time interval between successive releases of teeth on the timing gear—comes from some physical property, like the stiffness of a spring or the length of a pendulum. (As each tooth moves up a notch, it gives the timing mechanism a little kick to keep the pendulum swinging.) While clock owners must readjust the time every once in a while because of minor adjustment errors in the timing period, the hands themselves stay perfectly synchronized with each other.

The notion that a series of shafts could drive one another by means of meshed gears led to clocks with innumerable dials— representing the moon and planets as well as hours, days, months, and even years. Each new feature merely called for gears with the right number of teeth to get the desired ratio in the number of turns. For example, meshing a pair of gears with 25 and 75 teeth will cause the second gear's shaft to make one rotation for every three rotations of the first one. Furthermore, if one adds another gear with 25 teeth to the second shaft and meshes it with a 100-toothed gear on a third shaft, that third shaft will make one-quarter as many rotations as the second shaft (or one-twelfth as many rotations as the first one). Simple proportionality, that is, multiplication and division, governs meshed gear trains.

As gear technology improved, clocks and clockwork became fancier and more precise. For each design, someone had to calculate mathematically what would happen when the gears moved. Then in the seventeenth century a few mathematicians—notably Gottfried Leibniz in Germany and Blaise Pascal in France—began to explore the idea of turning the process around. Move the gears to determine what would happen in a mathematical calculation.

For example, imagine that I owned a 10-speed bicycle with a single 10-toothed gear on the rear axle and 10 gears attached to the pedal—one with 10 teeth, one with 20, and so on, all the way to 100. When the chain connects the 40-toothed pedal gear to the rear axle, each rotation of the pedal will cause the rear wheel to move four times. Then, if I wished to multiply, say, 4 times 7, I could count the number of rear wheel rotations caused by turning the pedal through seven full revolutions. In a real machine, of course, the shafts on both ends of my rudimentary multiplier would be connected to other mechanisms as part of a more elaborate calculation.

With hindsight, we might wonder how such a "simple" notion took so long to occur to anyone. In fact, gear-based calculating machines presented formidable technical design problems. They require components like ratchets, clutches, and differentials, not readily adaptable from existing clock technology. As a result, seventeenth- and eighteenth-century machines proved little more than intellectual curiosities, expensive to build and finicky to operate. Such machines stretched even twentieth-century gear technology. The mass-produced desk calculator I shared as a graduate student in the 1950s—a mechanical equivalent of today's cheapest hand-held electronic calculators—cost just about as much as my brand-new 1956 Chevrolet.

The precision machinery needed to provide the technology base for gear-based calculators began to emerge at the beginning of the nineteenth century. In 1822, a young Cambridge mathematician named Charles Babbage announced the invention of a machine capable of performing simple arithmetic calculations. But he never finished it. Instead, with the financial support he had obtained from the British government, he abandoned his original calculator and began work on a more ambitious design capable of handling more complex mathematical tasks. It also was soon dropped. Over the next ten years, Babbage began a series of designs but abandoned each one in favor of a more ambitious undertaking. Finally in 1833 he abandoned the calculator project completely in favor of a *programmable* machine. This forerunner of a modern computer was to be controlled by punched cards adapted from the devices French weavers used to control thread sequences in their looms.

Over the following thirty-seven years Babbage began a series of these programmable machines—each more complex than the last—and left each unfinished for the next in line. Babbage simply lacked the patience to follow any one of his ideas through to completion. As a result, his plans kept running ahead of his ability to produce the hardware he needed. In addition to a series of uncompleted machines, Babbage produced some fascinating speculations about future mechanical computation, including the possibility of "intelligent" information machinery.

As Babbage and his contemporaries realized, the ability to mechanize numerical operations also provided the means to manipulate non-numerical symbols. The designer merely assigns a numerical value to each keyboard character. Just like "modern" (mid-twentieth-century) mechanical desk calculators, Babbage's machines stored numbers in *registers* much like automobile odometers—a series of wheels, each marked with the digits from 0 to 9 and attached to ten-toothed gears. (Trip levers execute carries by kicking left-hand neighbors ahead one notch each time they pass zero.) By grouping the digits in such a register together, a designer can gain the flexibility needed to represent non-numerical symbols.

To illustrate what I have in mind, I'll create the following two-digit numbering system for alphanumeric characters. In my system, 00 through 09 will represent the digits from 0 to 9, while 10 through 35, and 36 through 61, will represent the lower- and uppercase alphabets, respectively, leaving 62 through 99 for things like punctuation marks and mathematical symbols. (Since my system devotes two digits—or a pair of ten-toothed wheels—to each character, the six digits of an automobile odometer can represent three-letter words. For instance, 121029 would spell "cat"—c=12, a=10, t=29.) In this way, *strings* of such characters provide a bridge between inherently numerical machines and representation of knowledge about the real world.

Driven by the excitement of such ideas, Babbage demanded too much from the mechanical technology of his time, and died with his dreams unfulfilled. Some years after his death, a combination of more precise gear-based machinery and less ambitious goals permitted others to build useful mechanical calculators that served as the workhorses of scientific as well as all commercial computing—consider cash registers, for example—through the first half of the twentieth century. By that time, clockwork technology had reached its ultimate limits and the next step in information machinery arose from an entirely different invention.

In 1825 an English inventor named William Sturgeon found that an electric current flowing through a coil of wire created a magnet. Shortly thereafter, the American physicist Joseph Henry discovered that placing an iron core inside the wire coil strengthened the effect—permitting this electromagnet to lift and drop small iron objects at the closing and opening of a switch connecting the coil with a storage battery. In 1837, another American, Samuel F. B. Morse, incorporated Henry's magnet into the first practical *telegraph,* separating the magnet from the switch by some five hundred yards of wire, and thereby demonstrating the feasibility of this method of "instantaneous" electrical communication.

When Morse pushed his switch, he closed the circuit. That action energized the distant magnet. When he released the switch, the circuit opened again, and the magnet returned to its quiescent state. The magnet operated a spring-loaded iron lever that produced the distinctive *click-clack* sounds that enabled telegraphers to hear messages sent by their distant colleagues.

With the switches, magnets, batteries, and connecting wire in place, Morse established a system for converting a set of electrical pulses, the "ons" and "offs" of the switch, into alphanumeric characters, which is the code that bears his name. Modern Morse code employs a four-state (or quaternary) system of dots, dashes, short spaces (between characters), and long spaces (to separate words from other words and punctuation marks). To minimize the number of dots and dashes required to send a message, Morse assigned the simplest combinations to the most common letters (one dot for an "e" and a dot and dash for an "a") leaving the more complex combinations for the less-used ones (three dots and a dash for "v").

For some forty years, the telegraph remained the last word in instantaneous communication over long distances. Then, in 1876 Alexander Graham Bell found a way of converting words directly into electrical current and back again, thereby avoiding the tedious coding and decoding associated with telegraphy. Instead of turning the current on and off in discrete steps, Bell's *telephone* caused the current to vary smoothly in proportion to the pressure created on a microphone by human speech.

While the telephone ultimately eclipsed the telegraph in most communications applications, the operation of the early telephone network made extensive use of "telegraphic" technology. In particular, routing calls required remote control of the switches involved, leading to the extensive use and perfection of *relays*—spring-loaded switches actuated by electric magnets.

With the growth of telephone service in the years following World War I, millions of telephones required interconnection. The demands placed on the telephone network of that period called for increasingly sophisticated telephone switches. That drove switching engineers to search for more powerful ways of exploiting relay technology.

For example, when I want to call the restaurant down the street from my office, the act of lifting my handset from its cradle alerts the local telephone switching system to expect a dialed number. In response, the switch connects my line to the specialized unit that will set up my call. That unit signals its readiness by sending me a dial tone. If I dial a local number, the unit arranges for a switchboard connection to the specified line. If successful, it connects me, adds a ringing signal, and gets itself ready to handle another call. If unsuccessful, it sends me a busy signal, connects me to a recorded message, or hands me off to a human operator, depending on the circumstances.

Long-distance calls require the participation of many switches. A call to Boulder, Colorado, from my Murray Hill, New Jersey, office, for example, would probably go from my local switch to a switching office in Netcong, New Jersey, for connection to AT&T's long-distance network. The Netcong switch might grab an unused line to Philadelphia and get that switch to connect it to a line to Chicago, and so on to Denver and, finally, Boulder. When I hang up at the end of the call, all those connections must be taken down in order to make them available for other calls. With so many remote connections required for telephony, it's hardly surprising that switch technology has always been a key R&D item in telecommunications.

In 1937 a Bell Labs engineer named George Stibitz recognized that electrical switches had a simple numerical property. In its "on" and "off" states, a switch can represent the digits 1 and 0. These two states constitute a *binary system* (in contrast to Morse's four-state, or quaternary, system, and the ten-state *decimal system* used to mark the pages of this book). Unlike Morse's arbitrary connection between his dots and dashes and the alphanumeric characters, the binary system represents numbers in an orderly way. The binary numbers from 0 to 12 go as follows: 0, 1, 10, 11, 100, 101, 110, 111, 1000, 1001, 1010, 1011, 1100. A register containing binary numbers must be about three times as long as its decimal equivalent, and binary computation calls for an awful lot of carries ($12+12=24$ in decimal becomes $1100+1100=11000$ in binary), but the natural fit to switch technology makes this approach very attractive.

Stibitz began by designing and building a *two-bit adder** (capable of adding just two binary digits) using a few pieces of wire, a couple of telephone relays, and a 1.5-volt dry-cell battery, all mounted on a slab of shelving board. This device produced an electrical output that corresponded to the sum of the electrically coded (1.5 volts for a 1 and no voltage for a 0) numbers on its two input wires. Since 2 is a two-digit number in the binary system (or $1 \mid 1 = 10$), the output required two wires. Placing 1.5 volts on both the inputs produced a voltage on the left-hand output wire (or 10), while 1.5 volts on only one of the inputs produced a voltage on the right-hand output wire (or 01).

Handling larger numbers called for replicating this device the appropriate number of times and wiring the combination together. Unlike Babbage's gear-based components, these electrical gadgets could accommodate any level of complexity that a designer might sketch. In contrast to the awkward way in which gear technology carried out numerical operations, Stibitz's† relays permitted designers to ship binary digits (or *bits* of information) to any desired location by merely calling for a connecting wire.

With this new technology, designers could explore new ideas without worrying about the limitations of "hardware." In es-

*John Tukey coined the term "bit" upon overhearing a conversation between two of his Bell Labs colleagues about the need to name "binary digits."

†Independent of Stibitz's work, other efforts pursued parallel paths in various places around the world. As each country was swept into the gathering storm of World War II much of this work turned toward military applications. As a result, many groups built their machines in isolation covered by a veil of wartime secrecy.

sence, they achieved an ability to build anything they could design and analyze. Fortunately, a bright MIT graduate student named Claude Shannon* uncovered a powerful kit of analytical "tools" at just the right moment.

Shannon recognized the equivalence of binary (1 or 0) systems and pairs of logical (true or false) states. The analysis of relay-based systems could therefore exploit an existing formalism of mathematical computation called *Boolean algebra,* invented in the middle of the nineteenth century by the Irish mathematician and logician George Boole.

Starting with the basic algebra most of us learned in high school, Boole added notation and rules for carrying out logical operations in addition to the mathematical ones, thereby combining mathematics and logic in a single system. By making the connection between Boolean algebra and relay-based machinery, Shannon catalyzed the creation of *digital logic circuitry.*

Stibitz's adder illustrates the logical properties of binary circuits. Imagine the two input wires and two output wires connected to four terminals labeled A through D, respectively. As I indicated above, a voltage on both A and B produces a voltage on C but not D—corresponding to $1 + 1 = 10$—while a voltage on either A or B (but not both) produces a voltage on D but not C—corresponding to $1 + 0$ (or $0 + 1$) $= 01$. If in place of 1 and 0 we now interpret a "hot" (connected to the battery) terminal as "true," the circuit actions we just described equate to "If A and B, then C" and "If A or B,† then D," respectively.

Conversely, we can create a two-bit adder out of two logic

---

*Shannon is better known as the founder of information theory, a subject he first brought to the scientific world's attention in his MIT master's thesis.
†But not *both.* This *exclusive* behavior is normally denoted by "xor" in logic circuits.

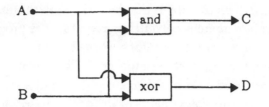

components, an "and-gate" and an "or-gate" (three-terminal devices whose respective output wires go "hot" when both or one of the inputs are activated). We merely connect one of the inputs of each gate to A and to B, the output of the and-gate to C, and the output of the or-gate to D. In numerical terms, therefore, a 1 at both A and B produces 1 and 0 at C and D, respectively, that is, $1 + 1 = 10$ as before.

This creation of digital logic provided the missing pieces needed to fulfill Charles Babbage's frustrated schemes of a hundred years earlier. It set the stage for John von Neumann's creation of the machine we now know as the modern computer.

Today

In today's world, the unmodified word "computer" normally refers to some version of the machine John von Neumann invented in the 1940s. In those days, a "computer" was a person who sat at a desk all day with a pencil, paper, and some kind of mechanical calculator. As more of the computation task shifted from people to machines, the name moved as well. While the term "computer" attached itself to a variety of machines, von Neumann's idea turned out to be so powerful that his invention quickly monopolized the word.

Earlier machines could store numbers and perform preselected sequences, or *programs,* of computing operations, but they lacked flexibility. Modifying any of these programs called for a tedious rewiring job. Instead, von Neumann created a

machine in which the stored numbers themselves could control
each operation, thereby permitting users to write programs and
ship them to the computer along with the data.

Like many other early computer designs, von Neumann's
notion envisioned a set of storage registers and a series of specialized processing components like the relay-based digital logic
elements described in the previous section—"adders," "multi-
pliers," and "comparators"—to perform the various computational operations. The computer also employed a network of
wires for moving inputs and results around the machine, as well
as a clock to provide a steady sequence of voltage pulses to keep
things moving—like the bass drum in a marching band. To
make control easier, each processing component was set up to
stay "off" unless energized by an external voltage.

Von Neumann's design added a new gadget called an *instruction decoder.* Picture a "box" with a register as its input and a
bunch of output wires. In its simplest form, each output wire
goes "hot" when the particular number assigned to it appears
in the input register. Sending the number 73, say, to this register
energizes a corresponding output wire. If that particular wire
were connected to the adder in such a way as to turn it on when
energized, then sending the *instruction* "73" to the instruction
decoder would cause the machine to execute the addition operation.

With each of the output wires connected to its corresponding
data-processing element, von Neumann's computer could be
operated by sending a sequence of numbers to the instruction
decoder. Because users could store these numbers just like data,
they could "rewire" von Neumann's machine by simply writing
a new set of numbers—or program of instructions—into the
memory.

Because the basic machine could deal only with numbers,
programmable computing machines were originally thought of

as no more than very powerful substitutes for mechanical desk calculators. In reality, however, they provide a flexible way of mechanizing the processing of the wide variety of symbols that *strings* (of alphanumeric characters) can represent. As we saw earlier, assigning numerical values to the letters of the alphabet gives the machine the ability to handle text. The amount of extra hardware required for this capability was so awkwardly large in pre-transistor days that it took people a while to wake up to that possibility. As advances in technology reduced the cost of hardware and increased its availability, people began to explore and appreciate efficacy of text manipulation. That launched a great wave of new computer use. At that point, computers became data processors—machines to help us deal with our records.

Giving computers the ability to handle text not only increased the kinds of data they could process, it also made them easier to use. In the early days, programs consisted of lists of numbers, each corresponding to a particular operation. Once a machine became capable of dealing with text, however, designers could arrange things so that a programmer could give the machine an "add" command by typing the three letters "a-d-d" on the keyboard.

In my earlier character representation example, I pictured a computer register as an array of two-digit decimal numbers, but real computers use groups of eight binary numbers called *bytes,* which encompass the numbers from 0 to 255 and amount to the same thing. This byte arrangement gives us room to represent all the letters of the alphabet and then some. As before, however, we will use 00 through 09 for the numbers 0 through 9, 10 through 35 for the lowercase alphabet, 36 through 61 for the uppercase alphabet, and 62 through 99 for things like punctuation marks. (Since every computer designer has many such opportunities for making arbitrary choices, most models of computers had a hard time communicating with others until the advent of industry-wide standards.)

In my system, the sequence 10,13,13 would spell "add." That means I would wire things up so that hitting the "a," "d," "d," and return keys on the keyboard would put 10,13,13 into some register. (Once made, such arrangements normally operate automatically.) In a real computer, I would also need a *compiler* program to transform the *command* "10,13,13" into the *instruction* "73" and send that latter number to the instruction decoder at the right moment in order to get an addition operation. If I do everything right, my machine will respond to the typed command "add" by carrying out that operation.

Today, the designer of a home computer might arrange things so that typing "hello" on the keyboard (thereby sending "17,14,21,21,24" to the central processor) would cause the display to read "Hello, I am your user-friendly computer. You and I can have lots of fun together as soon as you have memorized the first eight chapters of your easy-to-read instruction book."

To accomplish this trick, a program must connect the command "17,14,21,21,24" to a stored instruction. For our purposes, we can think of the computer's memory as a series of very long registers, each containing a line of text, and arranged in tabular form. Among other things, these registers must hold a list of all the commands in the computer's "vocabulary," together with the corresponding machine instructions that each command must trigger. The processor compares each typed input against its list of stored commands. Upon finding a match, it fetches the instructions stored with that command and proceeds to execute them—in our example, by sending the above-quoted message from its memory to the display screen.

While each step in the process is relatively simple, the results of their combined actions can easily seem like magic at first glance. As adults, we've had to learn to abandon many of the dreams of childhood, especially the yearning for magical power over the world around us. King Canute earned his place in British history by acting on his belief that the words that commanded his subjects would also subdue the waves of the English Channel. As modern adults, we may have given up on "abracadabra!" but many of us still shout "Come on seven!" from time to time. With the advent of software-driven machinery, the right set of "words" can give its user the ability to command and animate a lifeless pile of wires and switches. The right combination of character strings can even make their creator rich. (Like the few thousand words of Lotus 1-2-3, which yielded in excess of a quarter of a billion dollars.) If it isn't really magic, it comes close enough for some people to believe in it.

Magic or not, Von Neumann computers offer their users extraordinary amounts of information-processing power. This power has encouraged the use of logic-driven string manipulation to deal with all sorts of problems. Because computers deal in words and numbers, however, all information must be filtered through an alphanumeric format.

We humans encounter the same bottleneck in sharing our own experiences with others. Imagine a pair of tourists in the British Museum. Spotting an interesting piece of Greek pottery, one tourist might turn to the other and say, "That one looks nice." Those four words offer the second tourist little information beyond the object's location. But words can do more. In this case, John Keats's "Ode on a Grecian Urn" might help the visitor see an "unravish'd bride of quietness" sitting there on that shelf. Quite a difference.

The mechanisms that trigger a poet's ability to articulate ideas remain a mystery. Since psychologists know so little about how human minds transform the essence of an observation into words, we shouldn't be surprised that computer scientists haven't been able to "teach" that skill to their machines. How,

then, do we get help? For now, we can rely on computers for
help only with tasks that lend themselves to explicit description—like converting a page full of figures into an easy-to-read
spread sheet—and leave everything else to human beings.

Tomorrow?

As I look to the future, however, other alternatives appear
worth exploring. Personally, I find *neural networks* most in-
triguing. Instead of trying to make existing computers more
"brainlike" by feeding them the right instructions, some scientists have embarked on a totally new computing architecture in
which the circuit diagrams come straight from textbooks on
neuroanatomy.

The actual "computing" of the human brain has presented
scientists with a long-standing enigma. While studies of stroke
and accident victims have permitted neural scientists to associate certain neighborhoods of the brain with particular functions, most functions occur in more than one location and no
discrete analogs of "memory" or "central processor" appear
identifiable.

Tracing the component parts of even a small piece of human
brain is a staggeringly difficult job, requiring the painstaking
use of anatomical science's most powerful instruments. The
extent and intricacy of the wiring make the kind of "reverse
engineering" that electronics companies often do almost impossible in this case. A section of brain the size of a walnut might
contain over *ten miles* of wire—much of it too thin for viewing
with an ordinary microscope—made up of millions of pieces,
each of which is only a tiny fraction of an inch long. The entire
brain contains some ten billion individual neural cells, called
neurons, each connected to anywhere from hundreds to tens of
thousands of others.

By the early 1940s, a great deal of careful work had produced a rough "circuit diagram" of prototypical brain tissue. The individual neurons (brain cells) were known to become electrically active (or "fire") when stimulated through their multi-branched arrays of input wires (called *dendrites*). Upon firing, a neuron emits electrical signals over an array of output wires (called *axons*). These axons connect to the dendrites of other neurons (via connections called *synapses*), which either enhance or impede the firing of these other neurons whenever the first neuron fires. Since the firing of each neuron depends on the activity of many others, there is no easy way to figure out what ought to happen or when.

A few scientists—most notably two of von Neumann's contemporaries, Warren McCullough and Walter Pitts—managed to build simplified replicas of neural network out of electronic components, but no one could define a basis for understanding how the resulting circuit operated or what it might be good for. In the absence of promising experimental results, the pursuit of this approach gave way to the evident success of von Neumann computing. Except for a brief flurry of interest in the mid-sixties, most computer scientists regarded electronic neural networks as nothing more than an interesting historical oddity for some thirty years. Then, in 1982 a Bell Labs physicist named John Hopfield reawakened scientific interest in neural networks by finding a resemblance between their neighbor-pulling-neighbor structure and the behavior of magnetized atoms in some kinds of crystals.

To get any kind of useful result out of this analogy, Hopfield had to make a pair of simplifying assumptions about the "neurons" in his models—first, that his artificial neurons were either fully "off" or "on," and second that their synapses (connections) only helped to turn other neurons on. (Real neurons fire with a range of intensities and are able to discourage, as well as stimulate, firing action in others.) He also had to add an artificial general bias toward staying off, to keep neurons from all just flipping to the "on" state.

These principles in hand, Hopfield and his colleague David Tank now adapted the computational methods that simulated the behavior of flipping magnets to predict what ought to happen with neurons. They constructed a program designed to behave like a hypothetical *neural network* of one hundred interconnected neurons, but able to run on a conventional computer. Hopfield and Tank "wired" their simulated neural networks by entering or deleting coefficients in the mathematical equations the program used to keep track of each neuron's encouragement.

Depending on its particular arrangement of synapse connections, each simulated hundred-neuron neural network embodied its own set of possible equilibrium states (out of a total of one million trillion trillion, or 10^{30}, possibilities). Upon receiving an input word via its "dendrites," the system would settle into the most closely matching equilibrium word (a binary string of 1s and 0s represented by fired and unfired neurons) as each firing neuron affected the firing of others. Hopfield once described this process to me as "dropping a ball on a wavy surface and watching it settle into one of the valleys near its landing place."

Hopfield realized that each of his "valleys" or equilibrium words corresponded to a *stored word*—stored in the system by the act of selecting a synapse pattern. He and David Tank proved this point by using their simulated hundred-neuron network as a special kind of pattern-matching memory, known to computer scientists as "content addressable."

The procedure went as follows. A particular one-hundred-bit-long binary "word" (a string of 1s and 0s) imposed a firing pattern on the neurons, causing each neuron matched with a 1 to "fire," or turn on. Then the experimenters established a set of permanent synapse connections such that firing neurons provided encouragement to each of the other firing neurons—thereby creating an equilibrium for the imposed word. They repeated the process nine more times with different, randomly selected words—adding the required synapse connections to the ones already in the network from the earlier connections—resulting in a total of ten stored words.

With all the synapse connections in place, Hopfield and Tank presented the system with only a part of one of the words and left the rest blank. Nevertheless, as the system settled into a stable state, the entire word appeared as the final neuron pattern. It didn't matter which part they started with, the rest of it would emerge as the synapse connections from the firing neurons acted on the others. (The synapses associated with the other stored words just added randomly to the various inputs and did not affect the final result.)

Hopfield and his colleagues had used this neural model to create what computer scientists call a *content-addressable memory*. A conventional computer memory stores information like a filing cabinet. You have to remember where you put something if you hope to get it back. In a content-addressable memory (such as the human brain), on the other hand, you only have to remember some part of *what* you stored rather than *where* you stored it. If someone says "Dietrich" to me, my memory immediately finds "Marlene." I don't have to hunt through "actresses" or "famous names that begin with D" in some kind of alphabetical index inside my head.

As the fraction of the original word fed to the network decreases, a neural network sometimes ends up with a similar word instead of the right one. (For a long time I used to have trouble remembering the names of two fellow employees, Anne Anderson and Anne Alexander. If I came up with the wrong name first, I couldn't get to the right one.)

News of these encouraging results has awakened an explosive interest in neural networks and drawn people from a host of scientific disciplines. At Bell Labs alone, I can personally identify people with backgrounds in biology, chemistry, computer science, electrical engineering, linguistics, mathematics, physics, psychology, and zoology who are active in this field. In addition to computer simulations, the work now ranges from specially designed integrated circuits that incorporate several dozen neurons on a single "chip" to studies of the eating habits of common garden slugs.

New neural network circuit designs and new insights now permit attacks on a variety of problems, such as finding the shortest route that touches each of a given set of stops along the way. Instead of competing head-on with von Neumann computers, neural networks complement them. In a recent telephone directory experiment at Bell Labs, for example, a neural network took about as much time as a conventional computer to find the desired entry. It consistently outperformed the latter, however, in dealing with misspelled names. While the size of present-day neural networks limits them to just a few dozen names, future directory assistance systems may well offer both a letter-by-letter computer search for names the user knows exactly and a neural network pattern match for uncertain cases.

Looking to the future, researchers hope to create neural networks that they can program by presenting examples and encouraging correct responses, much the same way parents teach their children to identify "dogs" or "chairs." In the not-too-distant future, I expect such networks to find important application in the area of pattern recognition—for example, for speech understanding, monitoring engine sounds for bearing noises, and helping a robot identify objects in its field of view.

In such networks, each synapse must act as a quasi-independent entity, varying the strength of the connection at its location and attempting to correlate these variations with the success of the network as a whole. Early models attempt this task by using the equivalent of miniature von Neumann computers as synapse connectors. As a result, I think it's reasonable to expect future neural networks to take on many of the aspects of an interconnected network of conventional computers. To me, that suggests a possible avenue toward the synthesis between computation and biological brain function that modern technology has long promised but has yet to achieve. But that's not the only avenue, especially as science gives us increasingly powerful computing technology based on von Neumann architectures, as well as a host of other modern information technologies.

Modern
Technology

Every little bit counts.
—ANONYMOUS

THE FURNACE in my basement has a balky relay that gives me just enough trouble to remind me that it's there, but not enough to go through the hassle of finding a replacement. The relay's job is to turn on the furnace fan that drives hot air through my house. When the thermostat clicks, it sends a small electric current on a wire that leads to the basement and connects to a magnetic coil on my relay. The current flowing through the coil is supposed to (and usually does) get that magnet to pull the relay contacts together, thereby connecting the furnace fan to the power line. Looked at from the outside, a relay is just a current-operated switch. Relays range from the size of a lump of sugar to that of a breadbox, depending on the amount of current that needs switching. A transistor does the same thing, although usually on a much smaller scale.

Most transistors consist of a three-layered sandwich of *semiconducting crystals.* We can think of a transistor sandwich as two microscopically small slices of electrically conductive "bread" separated from each other by a thin slab of semiconducting "cheese."* The injection of a small amount of electric current converts the "cheese" from an insulator into a conductor of electricity. This small current acts to "connect" the two pieces of bread and allows a much larger current to flow between them.

When transistors first appeared, people spent a lot of time worrying about how to make them as big as relays and vacuum tubes so that they might someday serve as replacements for them. Then one day somebody inverted this perspective, pointing out that the transistor's ability to switch very small amounts of current—when commanded by even tinier inputs—constituted a unique property that people ought to look for ways of exploiting. The most important of these new uses was the switching of bits of information.

Lithography

As we saw in the previous chapter, the first computer switched bits of information with relays. Fortunately, we can now use microscopically small transistors instead, and that makes a fantastic difference. If I had to replace the fingernail-sized microprocessors in my compact disc player with equivalent relay-based computing circuits (containing switches like the one on my furnace fan) that unit would need a control box the size of a refrigerator.

Thanks to lithographic printing, the tiny transistors that microprocessors use to switch bits of information also have tiny price tags. A hundred of them would fit comfortably on the head of a pin and cost less than one cent. (A typical circuit

---

*The degree to which semiconductors conduct electricity or act as insulators depends on their composition. Crystal growers routinely control compositions to a few parts per billion. This control of impurities is as if New York City had achieved a degree of cleanliness such that no one but the mayor ever dropped a piece of trash on the ground.

might contain ten thousand transistors and cost about a dollar.) This affordability has allowed digital electronics to permeate every aspect of the modern world, from assuring the proper exposure setting on a compact camera to controlling air traffic.

Instead of making individual transistors one at a time and wiring them together, today's electronics manufacturers can print complete circuits—including the transistors—on thin sheets of semiconductor material. This lithographic (literally, "stone picture") printing process uses much of the same technology that produces the printed pages of our daily newspapers.

Printing replicates the raised features of an ink-coated "plate" by pressing this plate against a sheet of paper. Consequently, printing often begins with the removal of the light (ink-free) portions of the desired image from a flat surface. For many printing jobs, *photoetching* provides a chemical alternative to removing this material manually by "engraving" or *digging out* the light portions of the final image by hand. (The U.S. government's Bureau of Engraving and Printing still depends on this method to produce paper money.)

Photoetching was invented by Joseph Nicéphore Niepce of France early in the nineteenth century. (Niepce also invented photography. His partner, L. J. M. Daguerre, perfected Niepce's process and popularized *daguerreotypes* as the first commercial photographs.) Niepce oiled a paper print of an ordinary copper engraving—in order to make the ink-free portions of the sheet translucent—and laid it on top of an unmarked copper plate coated with bitumen, the gunky black tar that hardens into asphalt. After several hours of exposure to sunlight, the bitumen under the translucent portion of the mask hardened and became insoluble. This allowed Niepce to wash away only the unexposed portion of the bitumen coating with a kerosene-based solvent. He thereby obtained a negative image of the original drawing which he could then chemically engrave into the surface of the new plate—etching the unprotected metal with an acid solution.

Today *photolithography* (printing from photo-etched plates) not only produces the great majority of all plates for the printing industry, but also provides the basic technology for the electronics industry as well. In this latter application, layers of metal and other substances are deposited on sheets of semiconducting silicon crystals covered with *photoresist,* an updated version of Niepce's bitumen. The coated surface is illuminated through a mask (a cutout equivalent of Niepce's oiled paper) that casts a pattern, like sunlight streaming through a venetian blind. Once the light has caused the exposed areas of photoresist to harden, a solvent bath washes away the still-soft unexposed photoresist, leaving behind a pattern that replicates the holes in the mask.

Next, an acid (or other etching material that does not attack the photoresist) eats away the unprotected areas, after which the hardened photoresist is removed by yet another solvent (which dissolves photoresist but does not attack the material beneath it). As a result, the remaining material in the etched layer matches the pattern cutout of the mask.

Creating an integrated circuit on a blank slab of semiconductor involves several repetitions of this photolithographic process—separated from each other by the growth of new layers of material on the surface in order to construct semiconductor "sandwiches" and interconnecting strips of metallic conductors. Each of these materials can be applied by a variety of methods, such as electroplating, wetting the surface with a thin coat of molten material, or even "spray-painting." In each case the undesired material is etched away, and the process is repeated for the next layer.

When a transistor is needed at some point in the circuit, the designer creates slots in a mask.* The slots define the metallic layer so as to create the needed connecting wires. The designer also arranges for the right set of holes in other masks in the set to create a sandwich of the appropriate semiconducting materials at that spot.

Since the process produces all the transistors in a circuit at the same time and under the same conditions, careful control can produce a very uniform product. On the other hand, with just one dud in the circuit, all the other transistors become worthless. A circuit with 100,000 integrated transistors is a commonplace item in today's high-technology world, and the most tightly packed ones contain over ten million. Transistors can be made so small that a modern million-transistor chip could fit with room to spare on the face of a dime.†

Small size decreases the amount of material needed and lowers the cost. Low cost brings new uses, which in turn bring higher production volumes and even lower costs. As a result, integrated circuits have become the workhorses of the everyday world.

Sound reproduction is a good example. In Thomas Edison's original phonograph, the sound of the singer's voice caused a stiff piece of paper to vibrate. A needle attached to the paper transferred these vibrations to a soft wax record moving past

*Masks are far too tiny for handwork. Designers work at computer terminals, employing highly sophisticated versions of the "paint" programs offered by home computers. The resulting images are photographed and further shrunk by a system of reducing lenses, then "printed" photographically.

†That density has been doubling every twelve to eighteen months ever since integrated circuits were first invented, back in the late 1950s. If this progress continues for the rest of the twentieth century, some of my colleagues expect to see as many as one billion transistors on a single chip by the year 2000. At that point, the inherent "graininess" of semiconducting materials may well limit any further reduction in the size of individual devices.

it. The needle's vibrations caused the groove in the record to take on a wiggly shape—the analog of the variations in air pressure caused by the original sounds. After the recording was hardened, the process could be reversed. The groove's wiggle, as tracked by the needle, caused a corresponding motion in the attached piece of paper, which in turn recreated the sound in the air around it.

The invention of electronic means of amplification—first tubes and then transistors—improved the quality of the reproduction process. A microphone converted sound waves into an electrical analog that was amplified and sent to a magnet that either deflected a recording needle or magnetized a recording tape. Conversely, the recorded signal was converted back to its electrical analog and amplified to drive the cone of a loudspeaker. As the technology improved, reproducing the original sound faithfully (i.e., with "high fidelity") became easier and more commonplace.

Sound reproduction has become even better through the use of modern digital technology. Recording companies don't keep analog replicas of sound any more. Instead, they amplify the original signal just enough to permit its amplitude to be measured every few millionths of a second. A special-purpose electronic circuit called an *A-to-D converter* (A to D stands for "analog to digital") makes these measurements automatically and records the results as a set of numbers. Recording companies prefer to keep the "wiggles" in numerical form so that they can use digital electronics to handle them. Today the numbers can be purchased directly as dots on a compact disc.* Another

*Much of the growing popularity of compact discs springs from their almost total immunity to noise. Since the disc contains the data needed to recreate the sound, rather than a replica of the sound itself, minor imperfections make little difference. As a cookie cutter gets older, any nicks and dents it acquires show up in the cookies cut from it. On the other hand, cookie recipes in old cookbooks don't produce poorer cookies. Page creases and stains don't mat-

circuit, a *D-to-A converter,* in the compact disc player later converts these numbers back into analog form so the sound they represent can be amplified to drive the speaker.

Telecommunications companies are doing much the same thing. As I indicated in the story of my encounter with channel banks at the start of my Bell Labs career, more and more of our telephone calls are being converted into digits and sent down the line in that form for later reconversion into an analog signal at the far end. Huge quantities of "numbers" are involved. An ordinary three-minute telephone conversation calls for the transmission of more than ten million bits of information in each direction. Today's technology takes enormous numbers of bits in stride. I think it's fair to say that the technology of converting sound into digits and back again has become the technology of choice in such applications.

One of the most important benefits I look for from continuing advances in integrated circuit technology will be a dramatic increase in the amount and quality of pictorial information available on computers. Until recently, the size of memory usually limited what we could see on our computer terminals. The intensity of each displayed resolution element (or pixel) on the screen must be stored as a "group of bits" in the memory. When you multiply the number of possible pixels on a display screen by the number of bits per pixel* you begin to understand why few computer terminals were designed with anything approaching full graphics capability in the days of 64K (64,000

---

ter, as long as the key words remain legible. (At some point, of course, the words in the cookbook—and the digits on the CD's surface—become unintelligible. That's why I said *almost* total.)

*The number of bits varies with the number of gray levels and colors desired. A digitized black-and-white TV picture normally employs eight bits (or one byte) per pixel, for example.

bit*) memory chips. Today's high-resolution displays are setting a new standard of capability that helps to move us toward
the goal of picture-quality computer graphics.

It takes several million bits of memory to store a picture
without degrading its quality, but one-million-bit chips have
already become commonplace items, while four million and
even sixteen million bits on a single chip appears to be within
the capabilities of lithographic approaches expected in the early
1990s.

I remember when the first "massive" memory revolutionized
computing in the 1960s—more memory than many of us
thought we'd ever need, available at giveaway prices (about $8
per eight-bit word). It changed the nature of computing because
one could afford to do so many things so easily. For just a
million dollars in the 1960s, a national laboratory or corporate
computer center could buy 128,000 words of memory. In the
1990s, on the other hand, that amount of memory will be so
cheap that it won't be valuable enough to justify the cost of its
own little plastic package.

As integrated circuit technology capability keeps improving,
what was a roomful of equipment yesterday now fits into a
single cabinet, and I can imagine that today's cabinets will fit
on tomorrow's chips. One of the most important by-products
of this shrinking of technology is the individual ownership of
powerful data-processing facilities.

Just a few years ago, functions such as word processing,
financial analysis, and computer graphics required machinery
so expensive that it had to be shared by many users and centrally administered. Today, on the other hand, table-top computers provide their owners with all these functions—and many

*More precisely, 65,536 (or 2^{16}) bits.

others besides—thanks to the microchip revolution. Much as
the invention of movable type universalized the printed page,
the printing of electronic circuits has made the computer-on-a-
chip one of modern life's most ubiquitous commodities.

Microprocessors

As integrated circuit technology evolved, the number of transistors that could be "printed" on a chip grew from a few to
a few hundred and then to a few thousand. Somewhere in
the late 1960s the number became large enough to contemplate putting all the data-processing circuitry of an entire
computer on a single chip.* And so the microprocessor was
born.

In the early days of microprocessors, the amount of computing circuitry—as well as the size of registers—had to be kept
smaller than in full-sized machines. By the mid-1980s, however,
microprocessors had advanced to the same 32-bit word size in
their registers that the big machines use. The 32-bit word size
is very important. With a 16-bit word you can only handle
about 64,000 memory addresses at any one time (just as you can
only get ten million separate phone numbers out of seven decimal digits), which is why we must resort to area codes to
provide a unique number for each telephone in the United
States. With 32-bit words, the number of possible memory
addresses goes to more than four billion. The amount of addressable memory limits the kinds of programs, operating systems, and interfaces that a machine can handle.

Microprocessors with 8 and 16 bits make up for some of the
shortcomings of the smaller word size by segmenting large

*Only the computer's memory remained off-chip. Microprocessors almost
always supplement their on-chip memories with separate specialized memory
chips for economic reasons.

numbers into pieces and dealing with the pieces one at a time, at considerable sacrifice to their performance. For example, it's usually easy to spot personal computers built upon 32-bit microprocessors by their nicer-looking graphic displays. These 32-bit machines can afford to handle the large number of picture elements needed for a high-resolution image.

Today, only minor architectural differences separate the data-processing capabilities of top-of-the-line micros from mainframe computers. The most powerful of today's microprocessors perform like full-sized computers, offering multi-mip (millions of instructions per second) 32-bit machine capability from a chip no bigger than a fingernail—the so-called computer-on-a-chip. People still pay lots of money for big computers, but only because they can do other things besides crunch numbers, like provide efficient interfaces between large numbers of terminals and data storage systems. Even though microprocessors haven't quite matched mainframes yet, many designers are already moving toward the next challenge, building their own supercomputers by using multiple microprocessors.

This is a good time for new supercomputer ideas. While the power of our fastest single-processor machines continues to increase as technology advances, the rate of growth has slowed considerably in recent years. The interval between models with, say, twice the power of their predecessors seems to be getting steadily longer. As we approach the tail end of what we can squeeze out of single processors, most manufacturers of large computers now include two or more processors in their top-of-the-line machines.

"Big" computers rely on hand-wired processors that take advantage of high-speed transistors that don't lend themselves to microprocessor technology. But that's a moving target. Physicists, engineers, and materials scientists continually enhance the

performance of these devices, and create new ones as well. At
the same time, many of these same people are finding ways of
integrating these exotic technologies into existing materials systems.

Today's microprocessors use so-called CMOS (for "complementary metal oxide semiconductor") transistors, while their
hand-wired competitors use bipolar transistors.* Since the performance of the CMOS appears to be overtaking that of the
bipolar, some manufacturers of big computers are looking
beyond chips made of silicon for the next generation of hand-
wired machines. Some contemplate using gallium arsenide, an
expensive but widely used lightwave device material with attractive electronic properties. I see it as a horse race, but my
guess is that microprocessor technology will ultimately prove
the better bet.

At Bell Labs and elsewhere, not only are scientists devising
ways of providing our microprocessor designers with the benefits of bipolar technology, they have also made a promising start
toward the eventual merger of gallium arsenide with silicon. By
growing thin layers of gallium arsenide crystals on selected
portions of a silicon substrate we hope to "print" different kinds
of transistors in the same integrated circuit. That way, each
portion of a complex circuit may someday employ the exact
devices that best meet its particular needs.

Unless someone comes up with some radical improvement to
our technology that doesn't lend itself to integration, our best
hope for dramatic increases in supercomputer power lies in
learning how to harness machines with multiple microprocessor configurations. The biggest computers now usually come
with at least one extra processor in them for added perform-

*Both kinds employ silicon "sandwiches." CMOS produces more compact
devices, while bipolar tend to be faster and require more power.

ance. While today's supercomputers offer a lot more capability than any single microprocessor, the multiprocessor approach often allows micros to overcome their performance disadvantage by weight of sheer numbers. Designers can afford to use so many that they can produce virtually unlimited amounts of computing power for some applications.

Before the flood of hand-held electronic calculators changed the way school children compute, a billion computer instructions would have been enough to do one night's math homework for each of the elementary and high schools students in the entire United States. Today, the largest full-sized commercial computer would take about ten seconds to do a job of that size, while the most powerful 32-bit microprocessor would take between one and two minutes. (The smaller microprocessor most commonly used in personal computers would take about an hour.) Interestingly, the difference in power between the two is only a factor of 10. While the supercomputer is ten times more powerful, it can also cost one thousand times more money. So the obvious course is to link microprocessor-based machines together.

The need for really powerful computers becomes more apparent when we consider a more profligate use for our billion computer instructions than math homework. A billion computer instructions would have to be carried out for a computer to recognize a few seconds' worth of speaker-independent, connected human speech.* What a single human stenographer does in converting spoken utterances into written text would take several supercomputers working in perfect cooperation.

While devoting those costly machines to such a task is clearly out of the question, the same task could also be done by a

*Computers have an easier time matching isolated words from a single human speaker against stored examples.

few dozen high-performance microprocessors—provided that someone can devise a way of partitioning the job into small enough subtasks that each processor can shoulder an equal share of the load. Partitioning a problem to take advantage of multiple processors is crucial to the efficient use of such machines.

These days, it's hard to find a university in which some group in the electrical engineering or computer science department isn't planning to hook up a thousand or so microprocessors into a "supercomputer," telling people how many times more powerful than a Cray computer (the top-performing machines that have long set the standard for the industry) their particular machine will be.

At the same time, the multimillion-dollar machines that bear Seymour Cray's name are selling so well that universities, government agencies, and large corporations stand in line to buy them. The reason is that single very fast machines can usually execute a sequence of computational steps more quickly than a group of processors can separate them and eventually recombine them for a single result. Unless the task has some inherently independent parts—such as giving a roomful of customer-service clerks access to a large file of billing records—the communications overhead can easily grow cumbersome.

One can easily split up the job of providing computation for millions of homework assignments because they don't depend on one another. On the other hand, identifying the meaning of a spoken utterance provides fewer opportunities for partitioning. Unlike the neatly separated letters of printed text, spoken words and phrases blur into one another. This makes it difficult to devise and implement a recognition strategy that takes advantage of multiple processors.

In practical terms, most computing problems can be partitioned into a few large jobs and a number of small ones. As a result, general-purpose multiprocessor systems rarely manage to keep more than about half a dozen processors busy on any one job—one for each of the major tasks, a couple to handle all the minor ones, and another to keep track of the status of each subtask and coordinate their results.

While we have yet to devise a strategy for bringing the combined power of many processors to bear on problems like speech recognition, we know that efficient multiprocessing must be possible. After all, a typical New Yorker's speech-recognition system manages to orchestrate its multiplicity of neural processors to extract "why don't you" from "wine-tchah" with unselfconscious ease, even in the presence of traffic noises and other distractions of urban life.

On the hardware side, microprocessors and full-sized computers confront the same ultimate performance barriers, such as the speed of light, which limits the speed of electrical signals to one foot per nanosecond.* Nanoseconds count when information must move from one part of a computer to another. Hence the smaller the computer, the more rapidly its parts can communicate. In the final analysis, therefore, the distinction between microprocessors and "full-sized" computers will continue to blur as designers strive for the smallest possible designs.

Despite the explosive growth in computing, we have yet to feel the full impact of the information-processing resource that microprocessors appear capable of offering. As the "computer-on-a-chip" gives way to the "supercomputer-on-a-chip," technology promises unprecedented amounts of computing capacity at our disposal. This abundance will intensify the challenge that already confronts us, developing ever more powerful methods of telling our machines to do what we wish them to do.

*A nanosecond is a billionth of a second.

Software

Shortly after taking my present job as head of Bell Labs' research organization in late 1981, I asked one of my colleagues, Phil Anderson, to think about the most important long-term targets that the people in our organization might pursue. Phil is one of the world's great theoretical physicists,* so I naturally expected his suggestions to emphasize the physical sciences. But he surprised me. Instead of starting with "new methods for probing the limits of quantum mechanics" or "the study of individual atoms on crystalline surfaces" (both of which lay farther down on his list) he started off with: "Let's face it. Unless you solve the software problem, nothing else matters."

Solving "the software problem" means chasing a moving target, or a series of moving targets. It means producing software more quickly, with fewer bugs and at lower cost—software that's easier to understand, modify, and reuse in a different application. It means programs that give the user the power to customize a system by changing the software, while making sure that those changes don't cause that system to crash. Quite a list. While the items on the list don't change much from one year to the next, the "passing grade" in each category keeps rising all the time.

Today's software does miraculous things by the standards of earlier generations. Each of today's computers normally comes with a housekeeping program called an *operating system* that

*Among physicists, he is best known for his explanation of why "insulators" provide such perfect barriers to electric current. Ordinary pieces of glass or stone contain impurities which (it seemed) ought to provide a sufficient source of loose electric charges to conduct electricity. Anderson showed that these charges remain "localized" in the immediate neighborhood of the atoms from which they came. This "localization" theory helps guide the intentional introduction of impurities into semiconductors and earned Phil a share in the 1977 Nobel Prize in physics.

mediates between the bare machine and the user's programs. Most operating systems handle several programs at the same time, and keep them from bumping into each other. Among other things, the operating system oversees the translation of the user's commands into the nuts-and-bolts language of the processor and its peripheral gear—making sure that the letters "a-d-d" typed on the keyboard of a particular terminal will cause the computer to execute an addition operation on the relevant pieces of data. This permits the programmer to work with a higher level language than the one that drives the machine itself.

I learned how to program while at Columbia University in the late 1950s on an IBM 650. The latest word in computing in those days, the machine and its human helpers occupied most of a large brownstone building on West 116th Street and could do about as much as one of today's programmable hand-held calculators. Every time I moved a piece of datum, I had to specify both the old and new memory locations. In every mathematical operation, I had to specify the registers to be used for the operands and the result. Every time I wrote an instruction, I had to go through the pain of figuring out where a piece of datum was and how I could juggle the available registers in the processor to store intermediate results—making sure not to write over anything I was going to need later.

While some people were already thinking about operating systems and high-level computing languages, I had no access to such things. Instead, I kept laboriously detailed records of my data locations at each step. No matter how hard I tried, however, I would usually be forced to ask for help from the local professional programmers, one of whom took a particularly dim view of the clumsy efforts of graduate students and generally gave me a hard time.

Then one day, after I'd left Columbia and joined Bell Labs,

I learned about Fortran, a language in which I had only to write down the mathematical steps I wanted accomplished, and give each variable a name ("a," "b," and "c" were popular). With a "compiler" program to translate my program into instructions for the operating system, orchestrating arithmetic operations and keeping track of data locations became automatic services. I remember thinking (with considerable pleasure), "Nobody is ever going to need a programmer again, and that obnoxious fellow will have to look for a new job!" I figured that a language which eliminated the need for the bookkeeping that plagued my earlier experience would make programming so easy that all computer users would be willing—and able—to write all their own programs as they needed them. In reality, however, quite the opposite happened; the easier programming became, the more people wanted from each program and the more programs they wanted.

In the early days, computers were scarce, slow, and expensive. Designers of operating systems focused their primary attention on machine performance, making sure that the machine performed each of its functions as efficiently as possible. As a result, each machine's operating system took advantage of whatever shortcuts that particular hardware architecture offered. While this specialization generally permitted user programs to run faster, it meant tailoring each program to a particular machine and operating system.

In time, as computer manufacturers introduced new machines to replace their earlier models, they retained many of the machines' idiosyncrasies to save customers the trouble of rewriting software—and perhaps becoming tempted to switch to someone else's hardware. Thus the design approach adopted in the interest of retaining existing software ended up serving the interests of various computer vendors.

As advances in technology increased the amount of computing power available, computer researchers in the 1960s began to chafe under the cumbersome excess baggage from the past. At the same time, several major computer users launched drives to stem the rising cost of programming.

Bell Laboratories was involved in one such effort, an ambitious project called "Multics" undertaken jointly with General Electric and the Massachusetts Institute of Technology. As this work progressed, designers found themselves forced to make the system awkwardly large in order to meet the various requirements of the participants. While this complexity dampened much of the earlier enthusiasm for the concept, the project did lead to a commercially viable system employed by a small but devoted group of customers until the mid-1980s. In the final analysis, however, Multics' most lasting impact on computing came from stimulating two Bell Labs researchers, Ken Thompson and Dennis Ritchie, to produce a better answer, which became known as UNIX.*

In contrast to the Multics philosophy of meeting all the needs of multiple constituents, Thompson and Ritchie focused instead on a single goal—making the system convenient to use for themselves as programmers. They began by establishing a small but comprehensive set of principles for the system's behavior, which greatly simplified the programmers' interactions with underlying hardware. For example, most operating systems employ a variety of formats for moving information between a program and various entities such as disk drives, other programs, and other computers—something like using different kinds of stationery (and writing) for get-well notes and job applications. UNIX programs, on the other hand, communicate with each other and the rest of the world in only one way. This arrangement allows the programmer to plug programs

*UNIX is a trademark of AT&T Bell Laboratories.

together like the blocks in a Lego set, without worrying about the details of the hardware involved.

While numerous other features have contributed to the popularity of UNIX with programming professionals, its modularity, or building-block approach to software, has had the most far-reaching significance. UNIX's software modularity permits users to build customized programs out of modular parts from program libraries and programs borrowed from friends—together with the user's own specially written programs for the functions not covered by existing software. Furthermore, this modularity allows users to accomplish some quite remarkable things with just the standard utility programs that normally come with the UNIX system itself, as the following story demonstrates.

A few years ago, my friend Bob Martin was working at home one evening typing his wife's master's thesis in psychology. As he finished typing the final paragraph into his terminal, Bob noticed the repeated use of several unusual words. This observation led him to speculate about the use of word frequency as a tracer of subject matter. The notion intrigued him enough to devise a test. With the help of a computerized file management system Bob put together a small word-frequency program out of the standard UNIX text-processing modules. In short order, his program produced a list of all the words used in his wife's manuscript, along with the number of instances in descending order.*

Proud of the compact programming job he had done, Bob asked a number of friends to estimate the number of lines that it might take to write such a program. By the time I heard the story from him, he had "interviewed" several dozen people and

*His speculation proved correct. He observed that the top ten or so words (other than words like "the" and "are") would support an educated guess of the author's discipline.

categorized the results as follows: a senior executive of a large computing company turned in the highest estimate, twenty thousand lines; experienced programmers who didn't use UNIX came with much lower numbers, between a hundred and a thousand lines, depending on the language they normally used; finally, the UNIX programmers' estimates ranged around a dozen lines.

Once he finished telling me the results of his survey, Bob jotted down his version of the program on my blackboard with a big grin on his face. He had managed to do the job in a single line, consisting of the names of nine utility programs separated by vertical bars. In effect, the names of these programs had become the words of a higher level language, linked together in a one-line sentence.

While Bob's "one-line" program was judged remarkable enough to be featured in a 1985 story in *Fortune* magazine, it would hardly impress the users of today's most advanced word-processing programs. These people can get their systems to check word frequency in a piece of text by merely typing the appropriate command. Modern applications software for word processing, spread-sheet applications, and similar functions now offers user access to high-level commands—the ability to invoke a complex program-driven operation with a single word or group of words.

Access to the compact commands offered by higher-level languages helps to free users from concern with lower-level details. Nevertheless, the user must still specify the procedure to be used, that is, "make all letters lowercase, discard punctuation marks, alphabetize all the words in the file, count the number of lines occupied by the same word," etc., in the case of Bob Martin's word-frequency program.

While the creators of high-level languages try to make individual commands as convenient as possible for potential users, it's hard to anticipate individual needs and preferences. For instance, Michelé Velez takes care of the correspondence that flows through my office. We get several thousand letters and documents every year—many of which must be routed to others for their input and returned to me for action. Sometimes this process takes several iterations. Michelé must track all of those items and still have time to help out with other office functions.

In the summer of 1987, I got Michelé to try a new mail-tracking system that Al Aho, who directs our Computing Science Research Center, had created for use in his own office. Al shipped a copy of his program to our computer by electronic mail, and then dropped by to teach Michelé how to use it. While she listened politely and asked a few questions, I could see she wasn't all that pleased with the new system. Figuring she would like it better once she got used to it, I left her to work things out for herself.

Over the next few weeks, I'd occasionally check back and find that she was still sticking to the old system for one reason or another. My "improvement" idea wasn't getting very far. Then one day, I was surprised to find that the old system had disappeared. But Michelé hadn't finally learned to like the new one, as I had hoped. Instead she had rewritten the parts of the program that didn't suit her ideas of what such a system ought to do.

It seems that Michelé had enrolled in an introductory programming course that Bell Labs offers its nontechnical employees. Among other things, this course teaches the use of awk,* the high-level programming language Al had used to write the mail-tracking program. After only a few hours of instruction, Michelé realized that she knew enough to give the mail tracker some needed features, and quietly proceeded to add them.

*The creators of awk, Al Aho, Peter Weinberger, and Brian Kernighan, named their language after themselves, and wrote a book describing it (*The AWK Programming Language,* Addison-Wesley, Boston, 1988).

Thanks to some word-of-mouth advertising and encouragement from secretarial supervision, several hundred Bell Labs secretaries have adopted Michelé's version of Al's system. It's hard to tell which of the two is more pleased. Michelé got a lot of unexpected praise and attention for her creativity, and Al got a beautiful demonstration of the power of high-level languages—one of his pet causes in the computing community.

What's next? Just as high-level languages free programmers from worrying about details, some computer scientists are searching for a system of *nonprocedural programming* that will free users from worrying about *how* a given task is to be accomplished and allow them to merely state *what* they want instead (e.g., "I want a copy of this manuscript, together with comments from the people on the 'copy-to' list, as soon as all those comments have come back").

Given the amount of effort most of today's programmers must devote to the exact specification of procedures, the prospect of freedom from procedural programming can easily evoke the same feelings of "ultimate achievement" that freedom from the register allocation problem evoked in me a couple of decades ago. If and when nonprocedural programming becomes available, many present tasks will become easier and many presently "impossible" tasks will become feasible (e.g., "Reschedule all flights to minimize the impact of a twenty-minute takeoff delay at Newark Airport"). Nevertheless, if past experience is any guide at all, the technology of software production will remain hard put to keep up with the continuing growth of human needs and expectations.

Interfaces

In addition to providing users with access to nearby computing equipment, modern technology can also offer rapid access to remote sources of information and data processing. Just a few years ago, most people—including myself—thought of satellites as the ultimate means of global communications, powerful systems that could transmit tens, even hundreds, of millions of bits of data across continents and oceans in a single second. Today, however, I think it's fair to say that satellites have been eclipsed by the far more powerful communications capabilities of optical fibers. Each of these threads of glass can carry tens and probably hundreds of *billions* of bits of data each second. Furthermore, one can raise this enormous capacity even further by using a cable containing many individual fibers. This vast reservoir of transmission capacity appears capable of meeting all our data communications needs for the foreseeable future.

While light travels in a straight line through empty space, transparent materials can bend, or *refract,* its rays. In burning a hole in a piece of paper with a magnifying glass, for example, the parallel rays of a beam of sunlight are bent inward (toward the *axis* of the lens—an imaginary line drawn through the center of the lens, perpendicular to its surface) and converge on a single spot. Similarly, rays of sunlight traveling through a stack of such convex lenses would be continually bent inward toward the axis of the stack, thereby keeping the light moving through the central portion of each lens and preventing the light from straying out toward the edges.

While optical fibers are made of solid glass, the composition of the glass can be varied as a function of distance from the center in such a way as to give the fiber the same optical properties as a stack of convex lenses. Alternatively, changing the composition abruptly at a given radius creates a cylindrical *core* of glass within the fiber—so designed that rays of light within this core undergo total reflection at its outer surface. In both cases, light injected at one end flows down the center fiber and out the other end, even around bends.

In optical fiber communications systems, a laser light source is switched on and off rapidly, sending flashes of light down the fiber. In the most advanced systems now in use, each laser can transmit over a billion light pulses in less than one second, enough information-carrying capacity to transmit the contents of several sets of encyclopedias in that time. Moreover, light-wave systems engineers can combine a number of laser signals by means of a prism-like *diffraction grating* and send them simultaneously down the same strand of fiber. As long as each laser produces light at a different wavelength, another grating located at the far end of the fiber can split the different laser signals apart again. Early in 1985, Bell Labs researchers used this scheme to pack enough bits through a single fiber to accommodate a 1200-bit-per-second personal computer modem for every person in the United States over the age of ten.

The glass is so transparent that we need to beef up the signal only every fifty miles or so. If ocean water were as clear as the glass we use in fiber, a ship's passengers would have a clear view of the bottom for ten or twenty miles around—like a helicopter ride over the Grand Canyon on an even larger scale.

The simple fact is that technology is rapidly reducing the communications barriers imposed by distance to a state of near irrelevance. Today, physical barriers matter far less than the logical barriers that separate the sources and users of data. For example, today's technology could provide the president of a supermarket chain an instantaneous account of the firm's sales transactions, item by item, from the data collected by the bar-code readers now used at most checkout counters. But collecting and transmitting reports from thousands of cash registers across the country present less of a challenge than converting these data into a form that someone can use to make decisions.

As computers and software become more powerful and widely available, each new data-processing entity in the super-market chain becomes a potential source and user of information. Unlike their old-fashioned ancestors, computerized cash registers can make profitable use of electronic links to other systems. For example, the store manager's computer might monitor baked-goods sales throughout the day and drop the price on slow-moving items in order to empty the shelves by closing time.

Clearly, the automated on-the-spot pricing of baked goods helps both the store and its customers. At the same time, however, the aggregated sum of many such small-scale interdependencies complicates the lives of the senior managers charged with the welfare of the enterprise as a whole. To make sensible pricing, advertising, and resource allocation decisions, these managers need understandable access to the growing complexity surrounding their jobs.

To benefit from information created for different purposes under different conditions, users need convenient interfaces to the systems providing the data. Ultimately, the intervening networking technology that provides these interfaces should be flexible enough to accept information in whatever format the data source provides it—and should be powerful enough to convey the information in a format suited to the user's understanding.

Nature lavishes large amounts of processing power on the communications interfaces of living creatures. With current speech-recognition software, a computer must execute one billion computing steps just to identify the words contained in a few seconds' worth of normal human speech. A puppy can recognize speech—and much more—under environmental conditions that would drive acoustic engineers crazy. Human pattern-recognition skills, tactile sensitivities, and similar interfaces all attest to the massive processing power that our brains dedicate to such functions. I can't believe that nature has forgotten how to make simpler information processors. Instead, the experience of evolution has demonstrated the need for a variety of sensitive interfaces. Taking a cue from nature, I think that a great deal of the additional processing power now becoming available will find its best use in providing better interfaces between people and machines.

How soon will we be able to talk to our computers? Speech recognition won't necessarily give the machine any greater ability to deal with the meaning of spoken words than it now has with inputs from its keyboard. Since people normally speak to people and type at machines, the former method of communications usually results in more flexible responses. Today, the people I work with are looking for that same kind of conversational flexibility in our machines' interfaces—machines that can receive everyday voice or typed commands and respond with questions to resolve ambiguities.

While useful isolated-word recognizers already exist, getting computers to recognize normally paced connected speech presents a much more difficult challenge. Some engineers still regard speech recognition as a computing-power problem, a matter of waiting for big computers to get powerful enough to run existing recognition algorithms in real time. It seems to me that we can get results a lot faster if we are willing to look for clues beyond the speech waveform itself, especially through the use of syntactic and semantic modeling.

When people engage in conversation, they use their understanding of the context to fill in for the sounds they miss and for much of what the other participants leave unsaid. People make use of linguistic information to understand what the other person is saying. If we miss a word or two, we hardly even notice—as long as we are getting the content of the spoken message. If a voice-recognition system isn't able to understand English and keep track of content, we shouldn't expect as much from it as one that will.

The greatest subtlety of our own human interfaces appears in the way we seamlessly integrate disparate inputs. It's the single good feeling you can get in a theater from words, music, spectacle, and someone you like sitting next to you—all at the same time. In contrast, most of our present technology tends to deal with each input—the person, the words, the visual setting, etc.—as a separate entity. Worse, these systems can make little or no direct use of inputs that can't be made to fit the machine's character string–based format.

The personal computer workstation I want should be able to use direct human inputs such as spoken English as well as the things that people are used to interacting with, especially paper. Having a computer that won't handle unformatted paper is like having a housekeeper who refuses to enter messy rooms. We're so used to thinking of computers as powerful entities that we often overrate the level of difficulty of jobs the computer cannot do.

User preferences and productivity needs are the driving forces behind the call for better interfaces between people and machines. Computer manufacturers accustomed to grateful users and machine-oriented operators must now focus increasing attention on these interfaces. In the meantime, much of the burden of adapting to computer interfaces still falls on the user. Computers are advertised as "friendly" and "intelligent" while reluctant consumers are described as "computer phobic" or "keyboard averse." The message is that we ought to be grateful for what we're getting. Nonsense!

Personally, I'm "keyboard intolerant." I feel that the interface between machine and paper deserves much more attention. On top of the enormous amounts of paper records that exist from pre-computing days, almost every "computerized" organization generates ever-increasing amounts of paper that must be filed and accessed. To help meet this need, America's office equipment manufacturers produce about eighteen million filing cabinets each year. That's five brand-new empty cabinets for every newborn baby.

While nobody wants to have to rely on trips to the basement to dig through old files, most organizations continue to rely on antiquated recordkeeping methods and endure their consequences. Imagine an emergency patient rushed to the operating room of a large metropolitan hospital. An attendant is sent rushing to central files to pull the patient's record. In haste, the attendant finds the right aisle, searches along the cabinets, and then pulls the drawer out too far, spilling a handful of records on the floor. With someone's life in the balance, the only thing to do is grab the needed record and leave the others for "later." It's easy to picture what happens to such a file room as time goes on.

The advent of microfiche has produced a partial answer to this problem. Photographic images of several dozen sheets are mounted on cards, greatly reducing the space required to store the records and simplifying the job of keeping them in order. At the same time, access must be accomplished manually, making changes awkward, and browsing is frowned upon by administrators who must constantly worry about misplaced or mangled cards.

Many of the shortcomings of this system came from giving the designers too limited an assignment. Instead of merely replacing a filing *cabinet,* it's possible to create a new way of dealing with the information contained on the original sheets of paper by looking at what the user wants and working backwards. Microfiche helps reduce the bulk volume, but does little to make the material more accessible. Users need electronic access to high-resolution images of each sheet right at their desks.

In some offices, people can get an image of any filed "paper" record on the screen at their personal workstations in a matter of seconds. Hard copies can be made locally by a desktop printer, annotated if desired, and reentered into the system by feeding the marked-up sheet into the "copying" machine (which can also make paper copies if desired). "Copies" can be sent to others by sending the location of the stored record rather than a physical copy of the record itself.

To accomplish this, each workstation needs some additional interfacing hardware to mediate between the screen and the image-handling network. This network encompasses the computers that act as the image servers. The computers, in turn, control the means of image storage and retrieval.

The storage itself is usually accomplished by a combination of several methods (e.g., computer-controlled reels of microfilm, laser disks, and ordinary computer memories), each with its own advantages. High-resolution images (sharp enough that the small print on an insurance form can be read with a magnifying glass) of each sheet are stored either as digits by the computer or on reels of microfilm. In the latter case, the computer causes the microfilm storage units to fetch the appropriate reel, spool it to the desired frame, and create a digitized image for transmission to the user. With the help of a computerized file-management system, the user can have hands-on access to any of millions of documents in a matter of seconds.

Eventually, good interfaces to unformatted inputs (like scribbled notes and spoken words) will make our electronic helpers more useful, and may help to redress the imbalance that the computer's underlying tilt toward alphanumeric characters has imposed on us. Through the imaginative use of modern technology we can make computers more worthy of the extravagant claims made on their behalf by the various advertisers of "intelligent" machinery.

Chapter 6

Intelligence

Intelligence is the art of good guesswork.
—H. B. BARLOW, *The Oxford Companion to the Mind*

WITH COMPUTERS getting faster all the time, one might think that sheer speed could make human intelligence obsolete. Just get a computer to explore every aspect of every possible outcome to a given problem in order to pick the best course of action. While such a procedure would work in a tic-tac-toe game, the real world is far too complex for such brute force methods.

Exhaustive searches take more computer time than most people imagine. For example, let's take the old idea that gathering enough monkeys together and giving each monkey a typewriter to play with would result in one of them accidentally producing a valuable literary work. It's simply not true.

To see why, let me propose a much easier typing task, replicating the traditional typing sentence "Now is the time for all good men to come to the aid of their party." It comprises 68 characters, counting the spaces and the sentence-ending period. Since a typical typewriter has about 50 keys and two characters per key, each keystroke can have about 100 possible outcomes. Assuming that each new keystroke is independent of the previous one, three strokes can produce any one of one million

(100 to the 3rd power, or 100 times 100 times 100) different outcomes. Similarly, 68 random keystrokes would produce 100 to the 68th power, or 10 to the 136th power—written as a 1 followed by 136 zeros, or 10^{136}.

Compared to the original task, we've made the job much simpler. Let's also make the "monkey" more powerful. The most powerful of today's computers can execute one hundred million (100,000,000, or 10^8) instructions per second. Let's assume we know how to tie a billion (1,000,000,000, or 10^9) of these computers together to make a *really* fast machine. That way we would have a single "typist" who could produce one hundred million billion (or 10^{17}) different "sentences" every single second.* (Multiplying such numbers amounts to adding up their zeros, i.e., the numbers in the exponent.)

Next, let's get a lot of typists. A typical grain of sand contains a billion billion (or 10^{18}) atoms. Just think of the number of grains of sand there are on a short stretch of beach, and you get some idea of the enormous number of atoms our planet contains. If we move beyond the earth, to collect all the atoms in all the stars in our galaxy, the Milky Way, and then on to all the galaxies in the universe, astronomers estimate we could count a total of 10^{80} atoms.† Therefore, if we could turn each of the atoms in the entire universe into one of our typing supercomputers, we could get 10^{80} typists, or 10^{97} sentences per second.

Finally, let's give those typists "all the time in the world" to

---

*One new sentence per instruction greatly overestimates the capabilities of most computers. The supercomputer I have in mind would have to work on each character in the sentence simultaneously. The Cray computers mentioned in the previous chapter have this *single-instruction multiple data* (SIMD) capability.

†This calculation was first performed by Sir Arthur Eddington in the 1930s. Modern values of the astronomical constants give a slightly different value, but the method remains valid.

do their job. The universe has been around for some eighteen billion years. With thirty million seconds per year, that amounts to something less than 10^{18} seconds since time started. If our 10^{80} supercomputers had each been busily typing their 10^{17} different sentences each second from the moment of creation onward, we would have a total of 10^{115} (or 1 with 115 zeros after it) random sentences.

Remember, however, that we counted 10^{136} possibilities. That leaves us twenty-one zeros short. In other words, even if every atom in the universe had been working on the project since the beginning of time, we would have to speed up our "typists" by another factor of one thousand billion billion (or 10^{21}) just to get a reasonable chance of hitting the sentence as I originally typed it. Since most real situations offer a far greater number of possible choices than the keys on a keyboard, no wonder nature has given us a predeliction for seeking shortcuts to problems instead of blindly resorting to brute force.

Intelligent Behavior

The great computing science pioneer Alan Turing once said that he would call a machine "intelligent" when it held up one end of a dialogue well enough to convince him—via typed messages—that he was exchanging thoughts with a human being. In fact, however, it's a lot easier for a dumb machine to converse with a human being than Turing thought. People fill in missing pieces with unconscious ease.

In the early days of MIT's Artificial Intelligence Laboratory, Joseph Weizenbaum wrote a program called ELIZA to demonstrate how easy it was to meet Turing's test for "intelligence" by means of a few good guesses as to what people might expect from a conversation with a psychiatrist. For example, Weizenbaum expected a sentence fragment of the form "everybody . . . me" to reflect a recent experience. Thus, the statement "Everybody laughs at me" might elicit "Tell me, who laughed at you within the last few days?"*

When asked a question with a proper noun in it, for example, the program might respond with "Why are you interested in" followed by the proper noun and a question mark. Much like a nondirective psychiatrist, it often responded by inverting statements, or asking for more information about one of the nouns in the statement. Consider how "professional" a dialogue a random question can elicit from such rudimentary mechanisms.

Caller: "Are you really a psychiatrist?"
ELIZA: "Why are you interested in a psychiatrist?"
Caller: "I'm not really interested in a psychiatrist."
ELIZA: "You say that you are not really interested in a psychiatrist."
Caller: "That's right. I never had hang-ups about parents or stuff like that."
ELIZA: "Tell me some more about your parents."

Weizenbaum had provided ELIZA with a number of such response recipes. In addition, he took care to break things up by having the program select responses randomly and intersperse them with a series of encouraging stock phrases, like "Please go on."

Weizenbaum's program fooled people. In fact, it fooled them so well that for years afterward many who had conversed with ELIZA refused to believe that on the other end was a mere machine. When Weizenbaum finally pulled the program off the computer net, a great uproar ensued. Much to Weizenbaum's chagrin, a flock of MIT's computer users protested the end of their regular sessions with this friendly "therapist."

*Joseph Weizenbaum, *Computer Power and Human Reason*, W. H. Freeman, San Francisco, 1976.

Valentino Braitenberg's charming little book *Vehicles** debunks the notion that isolated acts of "human" behavior demonstrate a machine's intelligence. He does this by using the tools of electronics itself. Braitenberg's electronic actors are "vehicles," toy cars with a pair of light sensors in place of the headlights and a pair of small motors, each driving one rear wheel. The behavior of the wheels depends on how the motors and sensors are connected. In the simplest case, each light sensor drives the motor on its own side, so the more light a sensor gets, the faster the motor on its side of the car will go. As a result, these cars will turn away from any source of light they encounter. In a second group, the connections are crossed so that light on one side speeds up the wheels on the other. Since the wheels on the darker side always move faster, this model will always turn toward the brightest light source in its path, speeding up as it gets closer.

Braitenberg asks us to imagine a tabletop with some lighted candles on it. The first group of cars would turn and move away from the candles, while the second group would run around the table knocking down all the candles one after the other. Observing their behavior, he labels the first group "fearful," because they obviously slink off into a dark corner the first chance they get. By contrast, the second group is clearly "aggressive," attacking candles with systematic determination. With the addition of a few timers, voltage limiters, and the like, the list of labels begins to read like the table of contents of a Psychology 101 textbook.

For example, connect the sensors to the motors as in the first case, but add a switch that turns the motors off once a given intensity level is reached. In that case the vehicle would approach a candle, stop, and remain facing it—thereby demon-

*Vehicles: Experiments in Synthetic Psychology, MIT Press, Cambridge, Mass., 1984.

strating its "love" for the object of its undivided attention. My favorites include "egotism" (ignoring the candles and other vehicles), "special tastes," "foresight," and "optimism." Braitenberg's contraptions provide a neat demonstration of how easily one can believe in the "intelligence" of a few wires.

The point is basic. Confronted with isolated acts of "intelligent" behavior in cars—or with any "intelligent" entity—we must look deep enough to see how that entity functions under changing real-world conditions. If we don't, we cannot evaluate the true intelligence of a system—its ability to acquire and apply knowledge in a variety of contexts.

Representing Knowledge

As I see it, truly intelligent behavior calls for the ability to use information acquired in one situation to solve problems in another. Without that ability, a system is little more than a programmable calculator—driven by an exhaustive set of explicit instructions. Thus, the designer of an "intelligent" system must devise a *knowledge-representation scheme,* organizing information storage in a way that permits the system to apply that information in unforeseen situations.

An early "intelligent" program written at MIT dealt with mathematical word problems. At first, the program could recognize that the proposition "If farmer Bill owns six cows and buys three more, how many cows does he have altogether?" required the addition of two numbers. A later and more sophisticated version of the program even managed to convert the problem "If Tony is twice as old as Bill was three years ago and five years older than Bill is now, how old is Bill?" into a pair of algebraic equations and to solve them. The program's creator managed these neat tricks by setting down explicit procedures for sorting such problems into categories. For instance, he would use the presence of the words "how old" as the definitive identifier of "age" problems. In addition, he provided appropriate computational recipes for dealing with the problems in each category. Thus "was three years ago" would lead to the subtraction of three from Bill's age in the equation.

Once the performance level reached the level of high school algebra, however, it hit a wall—the intelligence wall. The program *was* able to deal with a variety of problems in various "contexts," but all of these contexts were sharply confined, rather as in a math textbook—straightforward and highly predictable.

By illustration, let's look at the following problem: "In 1985, farmer Bill and his wife Betty celebrated their twenty-fifth wedding anniversary together with their children—eighteen-year-old David, seventeen-year-old Jane, and ten-year-old Judy. In 1970, Bill and Betty had celebrated their tenth anniversary taking a trip to Disneyland with their children. How many people went to Disneyland on that 1970 trip?"

For that 1970 figure, the old MIT program would simply add three children to the two adults. A system with a more modern knowledge-representation scheme, however, might not make that mistake. A system today could, for instance, contain a rule that people get one year older every year, and that the system should deplete people with negative ages from its calculations. While human beings take such "commonsense" facts for granted, machines need explicit access to this kind of information. It's no accident that most of today's workers in the field of artificial intelligence (AI) regard the explicit representation of knowledge as their most important task.

The printed form, that burden of modern life, is in fact an important example of successful knowledge representation. For generations, this means of capturing information has joined rule-based systems—such as income tax laws, college admission requirements, or criteria for borrowing money from a bank—

and the real world. The chief attribute of the form is that it collects information covering *objects:* person; event; place; thing. For each object it lists a set of *attributes:* height; weight; age; prior employers; birthplace—for which it expects *values:* 69 inches; 178 pounds; 77; Artcraft Co., Henkar, Metropolitan; Munich.

Among attributes of Munich, we might find some that all cities possess, such as population, principal language, and location—as well as some unique ones, such as the world-famous *Hofbrauhaus.* Such lists provide a nice way of building a flexible knowledge structure. Designers of list-based knowledge-representation schemes associate a set of attributes with each object and record their values. Thus the system "expects" a person—but not a city—to have a certain "weight." Furthermore, since the value (Munich) of an attribute (birthplace) of one object (a person) is itself an object, linking lists provides a path for connecting information.

Suppose that such a system contained information about my father and needed to know what languages he spoke. If it had never recorded values for the language attribute, it could still check by hunting through the values of his other attributes (in this case his birthplace). One of them (Munich) would head a list containing information about language.

The flexibility of list-based systems matches nicely with the loosely structured way that information is often acquired. In my example, the computer might have gotten its personal information about my father long before it learned anything about Munich. Even though his personal profile said nothing about languages, the later link to another list would add that information without need for advance planning in setting up his original record, or any later correction.

Modern list-based systems normally contain general information about each class of objects, such as restrictions on the data-manipulation operations that may be performed on them. Thus the value of "age" normally advances by one year each passing year. A person's location can change, but not a city's. Numbers can't be added to colors ("six plus yellow" makes no sense) but colors can be added to each other under special rules ("blue plus yellow makes green").

For a long while, early AI researchers overstepped by using words like "intelligence" and "reasoning" as synonyms for the manipulation of list-based information structures. This blurring of traditional nomenclature is the source of much critical skepticism surrounding the literature of AI. Nevertheless, the manipulation of character strings (words, numbers, addresses, etc.) is the foundation of today's computing technology. As a result, lists of character strings and the links between them provide important building blocks for the representation of knowledge.

While much of the early progress in knowledge representation came from artificial-intelligence workers, few of the computer scientists who currently use such technology would attach the AI label to their own work. All designers of computer systems need to mirror the properties of real-world objects in abstract data structures regardless of terminology.

Once the properties of a particular area of interest have been adequately represented, designers must devise a strategy for moving the system toward its goal. We will next see how this combination might work in a well-studied example—the game of chess.

Chess

Ever since it was invented, the game of chess has been linked to human intelligence. It's a favorite pastime of scholars. The image of two people facing each other across a chess board, their minds deeply immersed in the game, is familiar to all of us. It's no wonder, then, that chess-playing computers appear intelligent to the casual observer.

In reality, however, the game of chess avoids many of the difficulties that computer programmers normally face in other problem areas. Consider the difference between describing the behavior of a white bishop and that of a beagle. The former can only occupy one of thirty-two chess-board squares, confined to the diagonals of its assigned color. The latter roams the neighborhood pursuing the appealing sights, sounds, or smells of the moment.

Unlike the real world, the world of a chess board—the locations and possible moves of each piece, as well as the pieces gained or lost as a result of each move—can be described without ambiguity. The clarity of the final-goal playing process is ready-made for computers. Introductory chess books usually help beginners learn the relative value of pieces through a point system—one point for a pawn, three for knights and bishops, five for rooks, and ten for the queen. In this system, therefore, each player begins with forty points and works to diminish the opponent's point score as the game progresses.

This numerical scoring method leads to a simple set of procedural rules—an algorithm—for winning chess games. Its only drawback is that using it is too tedious for even the world's fastest and most powerful computers. The algorithm is this: "Every time it's your turn, make a list of every possible move each of your pieces could legally make. Then, for each of these moves, list every possible countermove that you opponent can make. For each of these countermoves consider every counter-countermove and so on, until you've gone eight layers deep; that is, you've checked every possible counter-counter-counter-counter-counter-counter-counter-counter-counter-counter-counter-counter-counter-counter-countermove, keeping track of all the pieces lost by both sides in each possible sequence.

Finally, make the move on this list for which the least favorable outcome of all its possible consequences leads to the best net point win for you."

While such brute force wouldn't suit a human chess player, this recipe provides a traditional starting point for building chess-playing machines. But only the starting point. If a computer were to rely on this procedure alone, it would quickly be swamped by an explosive growth in the number of possible combinations. Until recently, chess-playing computers could analyze all possible outcomes from a given position for no more than four pieces (including the two kings), but can now handle five. With more than five pieces on the board, even the fastest of today's computers only analyze the consequences of every move and countermove to a depth of a few levels, never to the end of the game. Instead, designers invent ways of rejecting unprofitable sequences before they waste too much of the computer's time.

In the mid-1970s, two of my Bell Labs research colleagues, Joe Condon and Ken Thompson, designed and built a special-purpose computer—named Belle—to play chess.* Belle played a very good game indeed. As a result of a credible tournament record, "she" became the first nonhuman ever to gain an official master rating from the International Chess Federation. While Belle's level of play wasn't up to that of the world's best human players, it surpassed that of any other computer for a number of years.

Over time, however, Belle lost her edge as other computers continued to improve, while Joe and Ken devoted their attention to other computer projects. In the end, Belle was finally

---

*Outside the chess world, Ken is better known for his invention of the UNIX operating system I described in the last chapter. Belle's career is described in Jeremy Bernstein's book on Bell Labs, *Three Degrees Above Zero* (Charles Scribner's Sons, New York, 1984).

beaten by Crazy Blitz, a program that ran on the world's most powerful supercomputer.

What made Belle such a formidable player? First of all, Belle was wired to explore dozens of move sequences simultaneously through the use of multiple processors. As each move led to a group of possible countermoves—much like the decision tree for car repair I described in Chapter 3—processors were assigned to explore corresponding search paths. Since Joe and Ken knew the shape of the decision structure beforehand, they could optimize the layout of their processor network accordingly. As a result, Belle can evaluate over 100,000 positions per second.

Complementing this special-purpose hardware was a powerful set of "tree-pruning" rules that focused the search toward the most profitable possibilities, such as "If a move leads to a position which was previously reached by another move sequence, ignore it."

A chess player's goal is to checkmate the opponent's king. Naive programs, like poor players, focus on an obvious benchmark—relative number of pieces captured—and therefore often fall into traps. Belle, on the other hand, couldn't be tricked into giving away positional advantage in order to capture one of the opponent's pieces. Belle's moves were guided by strategic considerations (such as the importance of controlling the board's center, the value of doubled rooks on an open file, and the need to protect the king from attack) as well as by the value of pieces gained and lost.

Since neither Joe nor Ken was a chess expert, they sought the advice of people who were, and codified that advice in the rule structure that controlled Belle's move selections. Some of our more AI-oriented Bell Labs colleagues have characterized this rule structure as an expert system. In conversation, Ken admits to the merits of this notion—even though he rejected the idea when it was first presented some years ago—largely because the already fuzzy boundary between "AI" and "conventional" software keeps getting harder to identify.

During her tournament career, Belle played consistently better chess than the many thousands of intelligent human beings who play chess all their lives without ever achieving "master" status. Does that achievement constitute intelligent behavior? Personally, I would say no on two counts.

First, Belle explored hundreds of alternative sequences before making each move, most of which a human expert would recognize as unprofitable at a glance. Expert human chess players generally limit their explorations to a few possibilities. The blazingly fast speed of Belle's circuitry masks a relatively poor (i.e., compared to human) ability to tell good possibilities from bad ones. If human players had to explore as many alternatives as Belle does, a single game would take several centuries of around-the-clock play.*

Even though the rules of chess never change, the circumstances surrounding a particular game may offer opportunities for a change of strategy. At a recent Fredkiin match (an annual contest between a team of human chess players and an equal number of computers), a human player faced a computer that had just defeated another player whose ability matched his own. What to do? Recalling that the computer had been beaten

*Our experience with human problem solvers has led us to associate speed with intelligence. Smart people find shortcuts. On the other hand, that connection is less tenuous in electronic computing, which depends on the pace set by a clock. At each cycle, millions of logic gates move a bit of information from their input to their output. Increasing the clock speed doesn't increase the knowledge represented in the machine or the operations which the machine can perform on that information. While a computer's clock speed is a deciding factor in how fast it gets its work done, increased speed makes a computer only marginally smarter, at best. Speed offers nothing more than the ability to perform the same set of operations more often in a given time period (i.e., more "brute force").

in the first round of the tournament, our friend began his game with the move sequence that had led to the computer's earlier loss. To his delight, the computer responded exactly as it had before and lost in exactly the same way. While Belle wouldn't have fallen into that trap,* she has no way of adapting to un-expected opportunities the way humans do.

Flexibility

Chess's sharply defined problem boundaries and the certainty that the rules of the game will never change give brute-force computing methods an advantage rarely found in other areas of problem solving. As a result, designers of computerized prob-lem solvers strive to make other problems more "chess like" by imposing simplifying assumptions.

This is not a bad idea, this use of simplifying assumptions. After all, human problem solvers do it all the time. We physi-cists discuss "individual" atoms, "perfect" crystals, and "ideal" gases whenever we find their real counterparts too complex to handle. The trick is to do it well. As I'll show you with a few examples taken from the history of science, the quality of sim-plification separates intelligent problem solvers from the also-rans.

Artificial intelligence emerged as a struggling but hopeful discipline in the mid-1950s. From the beginning, one of its goals was the application of knowledge that would equal, or even exceed, that of human beings. After all, why should a million-dollar computer have any harder time recognizing an acorn in a forest than a chipmunk does? Furthermore, since many of the discrete acts people do—like adding up a bank balance or con-

---

*Like human chess players, computers rely on "book" openings, well-analyzed standard openings from the chess literature. Ken and Joe pro-grammed Belle to make random selections among the acceptable responses to each opening, and to avoid the ones that led to losses in earlier games.

trolling an oven—can also be accomplished by machines, intelligent performance might be a matter of connecting the right operations together. Over the years, however, that heady optimism has gradually subsided.

In the fall of 1978, I met the inventor of a program said to make scientific discoveries. The program had been fed the results of a series of experimental measurements on various gases, relating pressure, volume, and temperature. Sure enough, the program "discovered" that pressure and volume were inversely related. (Doubling the pressure on a sample of gas causes its volume to diminish by one-half—provided the temperature is held fixed.) This law was not unknown. Rather, the observation had been made in the middle of the seventeenth century by Robert Boyle,* and the relation between pressure and volume is called Boyle's law.

The computer program's inventor, very pleased, predicted that the program would soon discover many other laws of physics. The next step seemed to support such optimism. With a bit more fiddling, the program soon found Charles's law,† a similar relation involving temperature. (Doubling the temperature of a gas, at constant pressure, causes its volume to double. But that was it. Like the math-problem program, this one hit a wall. There was no way to extend the system to do more than find simple numerical relations between pairs of variables, so nothing more was heard from it, and little had been accomplished save to lead a computer to well-known conclusions.

The trouble with this approach reminded me of a public-speaking class I had at the City College of New York. As part of the course, each student had to teach the class how to do something. One of the people in the class chose "making a table

*Also by Edme Mariotte at about the same time.
†Discovered independently by Jacques Alexandre Charles and Joseph-Louis Gay-Lussac.

lamp out of found objects." After a flawless demonstration, the professor felt obliged to point out that the "found" objects happened to be the complete set of parts of a dismantled table lamp.

Herein lies a critical difference between the flexibility of human intelligence and its more rigid machine-based imitators. We remember Boyle and his colleagues for the *ideas* that led to the framing of these laws. Each of these scientists identified certain properties of gases that he thought to be related in interesting ways. Each pursued these ideas by designing and carrying out carefully planned experiments meant to reveal the behavior of these relationships. Their work hinged on the simplification that pressure and volume—or temperature and volume—merited an investigation that excluded all else. It was the selection of the problem space and the purposeful pursuit of its properties—out of myriad choices—that earned Boyle, Charles, and the others their places in history.

I have not come to the end of the story. Further investigations, carried out with better instruments in the eighteenth and nineteenth centuries, showed that the original workers had in fact "oversimplified" their problems. (Boyle and Mariotte assumed that the relation between pressure and volume was independent of temperature, as long as the temperature was held fixed. Conversely, Charles and Gay-Lussac held the pressure fixed and excluded it from consideration.) But there is a hitch. At low temperatures—ranging from $-40°C$ for carbon dioxide to $-253°C$ for hydrogen—the common gases become liquid, and Boyle's law doesn't work. Similarly, Charles's law doesn't apply at pressures high enough to press the molecules in the gas up against one another. Neither effect was observable with seventeenth-century equipment, so it was left to later generations of scientists to refine the work. That's how science makes progress.

Since these seventeenth-century scientists improved the understanding of gas properties in their day and pointed the way for later work by others, our hindsight could not label their work as "wrong." On the other hand, some of the world's greatest scientists have ignored relevant data and produced theories that were later abandoned. That's also part of the story of scientific progress, and of intelligence. We are a long way from machines. Intelligence permits humans to change the rules of the game when they get stuck. The following story points up that difference.

Lord Kelvin's work on thermodynamics made him a giant of nineteenth-century physics. In a time when thermal energy was promising to free muscle power from its age-old burdens, Kelvin's keen insight into the relationship between heat and other forms of energy won him a degree of respect that bordered on awe, both among his scientific colleagues and in the world at large. Thus, in 1862, when Kelvin presented his theoretical calculation of the energy storage capacity of the sun, everyone hailed it as an advance in human understanding.

A calculation of energy storage led to a calculation of the age of the sun—later called the "Kelvin time." Kelvin began by calculating the maximum amount of heat the sun—this body of gas—could possess and still be held together by the force of its own gravity. (He could not know that the sun has a way of replenishing its energy.) Using the relations between stored energy and the rate at which a hot body loses that energy through radiation, he demonstrated that the sun would cool to its present temperature in less than one hundred thousand years. It was generally accepted at that time that the entire solar system had been created in a single event; it followed therefore that the earth was also less than one hundred thousand years old.

This assertion put the world's geologists in a quandary. Their work had shown the earth to be many times that old.* One of them finally went to Kelvin and attempted to show him a piece of sedimentary rock, built up by the slow accumulation of debris at the bottom of some ancient ocean. From their measurements of contemporary sedimentation rates, the geologists calculated the age of their rock sample to be far older than Kelvin's one hundred thousand years. Kelvin didn't bat an eye. He coolly dismissed the hapless geologist with a famous one-liner: "There are two kinds of scientists—physicists . . . and stamp collectors."

For more than thirty years, these "stamp collectors," the geologists, continued their work, little noticed by their more famous colleagues. Then one day, another physicist dropped by and showed *them* a piece of rock. "I thought you might be interested in this bit," Ernest Rutherford remarked calmly. "I happen to know that it's over one billion years old."

How did Rutherford know this?

Rutherford, a pioneer in nuclear physics and the properties of radioactive substances, had discovered that pure uranium could be transmuted, that it could turn into lead—albeit at an incredibly slow but steady rate (about 1 percent of every uranium sample turns into lead in every one hundred thousand years). With this in hand, another physicist, R. J. Strutt,† had realized that the lead geologists always find in uranium ore was actually transmuted uranium. Since uranium and lead have

*Kelvin's result raised problems for biology as well, and led Thomas Huxley—Darwin's foremost proponent—to attempt a squeeze of the time scale for evolution into Kelvin's time scale.
†R. J. Strutt's notable contributions to science were overshadowed by those of his father, J. W. Strutt, one of history's most illustrious physicists. Like Kelvin, the elder Strutt was a baron, known to the world as Lord Rayleigh, a title that passed to his son upon his death in 1919. Kelvin, who was born with the name William Thompson, received his title in 1892.

very different chemical properties, their consistent pairing had been a long-standing puzzle. With Rutherford's discovery of radioactive decay, the ratio of lead to uranium in rock samples provided a reliable clock for measuring the ages of minerals.*

This was a great discovery. Great though it was, it put Rutherford in an awkward spot. He had to publicly declare Kelvin wrong and the geologists right—with Kelvin in the audience. Although a very old man by this time, Kelvin would certainly attend a lecture Rutherford was scheduled to give at Glasgow on his radioactivity work. Mineral dating had to be mentioned, but how to ease the blow? Fortunately, Rutherford found a way to avoid attacking Kelvin on his home ground by proposing an entirely new source for the sun's energy, one much longer lived—energy from nuclear radioactivity.

In fact, however, Rutherford's surmise about the sun's radioactivity was only partly right. While nuclear power fuels the sun's energy, the process involves the fusion of light nuclei (hydrogen into helium) rather than the disintegration of the heavy ones (such as uranium) that Rutherford studied. While the whole answer didn't come out until much later, Rutherford's insights put the subject on the right track.

On the other hand, we shouldn't judge Kelvin too harshly for dismissing the geologists earlier in the game. He had to choose between ignoring evidence from an unfamiliar source or destroying the simple beauty of his theory. After all, thermodynamics was a very strong branch of science at the time. It was known to describe the behavior of candles, steam engines, and volcanos. Kelvin's judgment rejected the idea that a law of

---

*Transmutation occurs in every radioactive element, albeit at widely differing rates. For example, the transmutation of a radioactive isotope of carbon (carbon-14) into nitrogen is an important tool for establishing the age of archeological specimens.

nature would contain a special escape clause for the sun. The problem was finally resolved, not by modifying thermodynamics, but by adding nuclear physics.

On the day of the lecture, Rutherford rose and began, "Some years ago Lord Kelvin demonstrated that, *without some new source of energy,* the sun could only have existed for a mere one hundred thousand years. My subject today is the *experimental verification* of that far-sighted prediction." Settling comfortably in his accustomed chair, Kelvin smiled contentedly and dozed off through the rest of the lecture.

Rutherford's inventive reinterpretation of Kelvin's earlier work thus solved his problem, and gave us also a fine example of intelligence in action. In the course of this story, Rutherford's problem-solving abilities moved smoothly from one context to the next. Beginning with nuclear physics, he applied his results to geological samples, then on to astrophysics, thermodynamics, and academic politics. Our machines, on the other hand (ELIZA, Braitenberg's "vehicles," various expert systems, and even Belle, that gallant queen of the chess board), each failed to operate in any but its intended context, and thereby failed to demonstrate true intelligence.

We humans live in an unpredictable world. While unexpected events sometimes give us pause, new situations also offer us opportunities for growth by allowing us to see familiar concepts in a new light—and thereby create new ideas.

Chapter 7

Ideas

Did you ask any good questions today, Isaac?
—JENNIE TEIG RABI

ISAAC ISADOR RABI, one of the twentieth century's most distinguished physicists,* invented a sensitive technique for probing the structure of atoms and molecules and thereby opened a fruitful new field of science in the 1930s. That achievement, and many others over half a century, earned Rabi every significant honor a physicist might hope to win. When an interviewer asked Rabi to speculate on the reasons for his success, he replied with the above quote from his childhood, his mother's habitual greeting when he returned home from school each day.

I got my postgraduate education at Columbia, where Rabi was a member of the faculty. In those years, I hardly knew him, even though he taught one of my courses. Our personal relationship developed some twenty years later, when we began to meet at the various social functions that bring Nobel Prize winners together in the New York area. We got on very well. He treated me like a fresh young squirt, and I enjoyed acting the part around him.

"How the other kids must have hated you in school,"

*Sadly, I. I. Rabi died in January 1988, a few months before his ninetieth birthday.

I ventured during one of our exchanges. "A well-scrubbed, well-prepared kid who knew all the answers and just wanted to see how well everyone else was doing." "Not at all," he countered. "There are questions which illuminate, and there are those that destroy. I was always taught to ask the first kind."

"Questions which illuminate" help nourish ideas. Ideas build knowledge. Students of any age need to nurture the kind of freely inquisitive spirit that Jennie Rabi encouraged in her son.

Have you asked any good questions today?

Asking Questions

Every small child I know appears to be born curious. Young children experiment with everything they can reach. They usually ask questions from the moment they learn to talk. Unfortunately, the flow of questions often slows down once they start school—unless they get the kind of encouragement that Isaac Rabi had. Personally, I can think of no better head start to healthy use of technology than a well-developed habit of inquiry. For adults, this habit is hard to maintain, for while truly dumb people often live in blissful unawareness of the gaps in their understanding, most intelligent people secretly suspect that their personal gaps are large—far larger than the people around them realize.

My first boss at Bell Labs—and one of the smartest people I have ever known—once confided a terrible secret: he felt overrated. As he spoke, I realized we shared exactly the same feeling. I could readily picture me saying those same things about myself. "The people around me think I'm smarter than I really am . . . They don't suspect that I learn things more slowly than they do and less fundamentally . . . When I listen to a presentation on something new, the only thing that keeps me from appearing stupid is to not ask questions about things that others obviously already understand (I know they understand it because they're not asking any questions) . . . The things I know about are really easier to understand, almost simple by comparison to my colleagues' areas of expertise." Since then, I've found that such monologues are common indeed.

What happens when such a secretly insecure person explains something to another with the same secret self-image (or, worse yet, to a group of them)? Since A has labeled what he understands as "simple" by self-definition, A can't insult B's intelligence by going into excessive detail. A's original estimate is reinforced by the obvious ease with which B appears to take in the material—nodding occasionally and never asking for additional clarification. Accordingly, A speeds things up even more, with no change whatever in B's demeanor. Off B goes at the end of the meeting, resolved to learn what A was transmitting during the exchange, wondering if A suspected his lack of understanding.

In my experience, the attitudes that underlie this wasteful undercommunication between technical colleagues often carry over into interactions between technical people and the rest of society. If engineers and scientists fear a public display of their "ignorance," of course nontechnical people rarely question technical presentations enough to get the information they need.

Many technical meetings are saved by those who feel free enough to ask questions. When a question is asked it's easy to spot at least half a dozen heads going up, others who would like to know the answer but couldn't ask for themselves.

In discussing this subject during a recent college lecture, I mentioned my daughter Mindy's asking how flipping a car's rearview mirror produced a new image. I went on to recount my explanation, as well as her response. "I'm really glad I asked," she said. "I thought I was the only one in the world who didn't know how it worked."

But the story didn't end there. It took an unexpected twist when I asked one of my Bell Labs colleagues to look over an edited transcript of that lecture prior to its publication. His marked-up copy came back to me with a scribbled note: "I'm not sure your flip mirror explanation is correct."

Perplexed, I called another Bell Labs physicist, and she told me that automobile companies make rearview mirrors by simply putting the reflecting surface on a wedge-shaped piece of glass. While most of the light travels through the glass, hits the reflecting surface, and bounces back—just as in an ordinary mirror—a small fraction reflects off the front surface of the glass, thereby creating a second (weaker) mirror which is tilted with respect to the first one.

I had mistakenly imagined that the manufacturer ruled lines on the back surface to create a diffraction grating—creating the same effect that sometimes produces a second (weaker) rainbow inside a larger one.

Sometimes we don't know what we don't know until someone asks the question. My daughter's ignorance of the technology in this case was no deeper than mine.

Tools

Technology is clearly improvable. It's a safe bet that tomorrow's technology will be far more powerful than today's. But people can improve, too, and people have also much room for growth. As we review the needs and opportunities for advanced technology, we should also look for growth in human capabilities. Some of that growth can come from better use of the tools technology provides.

Will the next generation be better equipped to handle technology than the present one? Like most communities, the town I live in has added "computing" to its high school curriculum. As a result, many of our high school students have learned to write elementary computer programs. Dealing with a keyboard teaches a useful basic skill, just as getting behind the wheel of a car teaches safe driving. Both skills help users "handle" technology in similar ways. They teach people to use existing tools.

Some young people learn to make imaginative use of computing tools long before they enter high school. I first met Dahlia Schwartz when she was still in second grade. Her grandmother—Lillian Schwartz, a pioneer in the computer art field, and a longtime Bell Labs consultant in computer graphics—brought her to my office for a visit. At the time, Dahlia was already quite comfortable with computers. She routinely used the word-processing program on her father's home computer for her letters, various games, and school assignments. Moreover, on trips to her grandparents' home, she frequently got an opportunity to explore computer graphics on some of her grandmother's more powerful machines.

During Dahlia's visit to my office, much of our conversation centered around the special features of my personal computer terminal—such as an automatic dialer which could extract phone numbers from my electronic mail messages and dial them at the touch of a button. It was a delightful visit and, as she left, I told her that I was looking forward to seeing her again soon.

Dahlia's next visit to her grandparents' house began on a Friday afternoon a few months later. While a couple of activities had been scheduled for the weekend, Dahlia suggested adding a Saturday trip to Bell Labs. Lillian tried to discourage the notion by pointing out that a visit outside normal working hours would take special permission. "Oh, I don't think that will be a problem, Grandma," the technically minded nine-year-old replied. She just sat herself down at Grandma's modem-equipped home terminal and began pressing the appropriate keys. "Arno Penzias must be able to give permission," she said. "I'll just send him mail and see."* If Dahlia's grandparents hadn't talked her out of it, Dahlia's imaginative use of her computer skills would have yielded the permission she sought in under five minutes.

Learning to program a computer is not an end in itself. Some of the most productive computer scientists I know never write programs. Rather, they work on underlying principles—such as the theory of algorithms. Algorithmic design is anything but rote; it calls for creative ideas rather than familiarity with computer machinery. For example, suppose I were to ask you to create an algorithm, namely: devise a method for finding all the anagrams in a piece of text (like "deal" and "lead" in "Can cane sugar lead to a good deal of acne?"). While a brute-force approach could get the job done on one page of text, only a sophisticated algorithm, working billions of times faster than brute force, would reveal the anagrams in something as large as a book. How would you do it?

If your plan consisted of matching the first word, letter by letter, against the letters in each of the words that followed it—and then did the same with the second word and so on—you can at least give yourself credit for understanding the problem. Comparing each word only to others of equal length is a further step in the right direction. But it would be better first to sort the words by length, so that words only needed to be compared within smaller groups. That's worth a part score. A simple way to keep from testing the same word more than once is to alphabetize all the words before starting the comparisons.

Jon Bentley, a friend of mine who teaches programming and writes about it, includes such examples in his courses to emphasize the need to back away from the machine and think creatively. Jon's answer to the above problem is *first* to alphabetize the letters in each word, and *then* alphabetize the resulting

*Like most regular computer users, Dahlia uses the unmodified word "mail" to mean the electronic kind sent over computer networks.

words. In that case the sentence "Can cane sugar lead to a good deal of acne?" would read "acn(1) acen(2) agrsu(3) adel(4) ot(5) a(6) dgoo(7) adel(8) fo(9) acen(10)" after the first alphabetization and "a(6) acen(2) acen(10) acn(1) adel(4) adel(8) agrsu(3) dgoo(7) fo(9) ot(5)" after the second. Clearly words (2) and (10), as well as (4) and (8), form anagram pairs.

Alphabetizing the letters in each word makes the letters they contain easy to compare. That's the "trick."

In my view, the ability to find such creative solutions comes closer to the kind of "computer literacy" that people need in a high-tech world than does sitting at a keyboard memorizing commands.

Building Ideas

A magazine interviewer once described me as "endlessly talkative." Most of my friends would agree with that assessment. I use words—in large numbers—to sift ideas and try them out on others.

On the other hand, a single well-chosen word can sometimes create an inspired image that can transform a familiar object into a new concept. When Akiro Morita led Sony into the production of transistorized radios, he wanted something more exciting than "the world's smallest portable radio." After some thought, Morita challenged his engineers to create a "pocketable" radio.

In practice, the early Sony models needed help from a tailor who provided Sony salesmen with oversized shirt pockets when the standard size proved a bit too snug for these radios. Nevertheless, the name "pocketable" not only gave designers a clear picture of Morita's vision, it also captured the imagination of the public. While Morita never said *which* pocket his radios would fit, his beautifully simple expression succeeded in getting his idea across to others.

There are more ways of combining words into grammatically correct English sentences than there are atoms in the universe. With so many possibilities to choose from, I see no disgrace in not finding the "right" one the first time. Moreover, a "bad" idea might be the first step to a better one.

The late Rudi Kompfner had more "bad" ideas than anyone else I've ever met. When I joined his radio research laboratory in 1961, Rudi was already one of the world's most celebrated electrical engineers. With a number of important inventions to his credit, he personally created several of the key elements that made satellite communications possible. But most of his ideas seemed hopelessly naive.

Rudi popped into my office one day with his latest "invention." At the time, our laboratory had been exploring a new communications network based on compact radio transmitters and receivers mounted on aluminum poles—somewhat larger versions of the ones used for highway lighting. But the system had a flaw. Whenever the poles swayed in the wind, the carefully aligned antennas would miss their intended targets. Rudi had "solved" the problem with a complicated rig for stabilizing the antennas atop their swaying poles. While I had to admit that his contraption could accomplish the job, I couldn't imagine that anyone would put up with all the extra struts and guy wires it called for. They took up too much real estate. There had to be a better way.

After we had gone back and forth exploring alternatives for a while, I hit upon the idea of putting the stabilizing wires inside the hollow pole itself. I was pretty sure I had something, but I knew it would take some work. Rudi said that was fine with him and promised to chat again in a few days.

By the time we met again, I had completed my design and even produced a small model of the tower—with a flashlight standing in for the antenna's radio beam—to prove my point.

When Rudi gave my "tower" an energetic push, the motion caused my wires to rotate a pulley connected to the flashlight just enough to keep it level. Rudi was clearly delighted to see the flashlight's beam cast a steady spot on the far wall of my lab. "Good," he said, "now I can go think about something else." Rudi's "bad" idea had produced my first patent.

Before the German "anschluss" forced Rudi to leave his native Vienna in the thirties, he had worked as an architect. His first day on the job, his new boss showed Rudi to a desk, gave him a map of a suburban building lot, and told him to design a two-bedroom house with a certain total floor space. That was it. Rudi was left on his own to carry out the task.

Stunned by the enormity of the task and afraid to put pencil to paper, Rudi sat immobilized. After a while, his boss returned, saw he was having trouble getting started, and offered to help. "Just put a square house right by the road," he said. "That way, they'll have a big backyard and won't have to shovel snow off their walk in winter." Rudi saw that his boss's plan wouldn't work. The people would want the privacy and appearance benefits a front yard offered. Furthermore, an L-shaped house would suit the triangular geometry of the lot. Rudi sketched what he had in mind.

"Fine," his boss said. "Let's have the front door open right into the kitchen. That will make bringing in the groceries so much easier." Again Rudi objected, and pointed out a good place for a side entrance that could lead to a more conveniently located kitchen. By the time the exchange had moved to a third "suggestion," Rudi caught on to what his boss had been doing—verbalizing an idea, *any* idea, to get the creative process started.

When Rudi retired from Bell Labs, he moved on to Stanford University, where he taught a course in experimental design. One of his classes designed a wind-powered generator, which required a wind tunnel for operation. No wind tunnel being available, Rudi offered his ancient convertible in the service of research. His students mounted their models on the trunk lid, and with Rudi at the wheel, the group conducted their wind tests by driving around the Palo Alto campus as fast as the campus police would permit. While Rudi's class never succeeded in building a commercially viable windmill, his students certainly learned how to build ideas.

Estimates

Ideas flow better with needed facts at our fingertips—fewer interruptions for look-ups, and fewer detours down blind alleys. Fortunately, everyday experience has given each of us a large store of relevant information—much larger, in fact, than most of us suspect.

Do you know how fast the Mississippi River flows? No? Suppose I told you it flowed at eighty miles an hour, would you believe me? Just picture an old paddle wheeler floating gently along with the current. Surely it isn't outdistancing a speeding automobile. Even though you didn't know the speed to three decimal places, your prior knowledge let you make an *estimate* that showed my wild assertion to be invalid.

A very high barrier stands between us and the habit of making rough estimates—the fear of getting the "wrong" answer. Contrary to what most of us learned in school, however, an inexact answer is almost always good enough. All through elementary school, high school, and the first two years of college, I was taught that only the exact answer would do—"July 14, 1789"; "5,280 feet"; "r-h-o-d-o-d-e-n-d-r-o-n." It was like a high-wire act. The slightest imbalance would send everything tumbling downward toward an inadequate safety net called "partial credit." I well remember the moment when the spell of that attitude was broken.

Henry Semat, my college atomic physics instructor, had completed a calculation on the blackboard, and several of us thought we had caught him in a "mistake." His answer was almost twice as big as the one given in the book—and he had written the book! Nevertheless, he didn't bat an eye when we pointed out the discrepancy to him. "Same order of magnitude," he shrugged with an air of total unconcern. "As long as we know that the effect is the right size, we can always fill in better numbers if we need to."

It was a revelation to find that a real-life practitioner of my intended profession didn't feel obligated to fill in every decimal place. An important lesson. Keep your eye on the left-hand digit and put zeros in all the other places. If you don't know the first digit, make a rough estimate and pick one.

I recently needed to know something about the long-term effect of radioactivity on transoceanic optical fibers. Cosmic rays flash regularly into our atmosphere from high-energy nuclear reactions, the product of massive stellar explosions within our galaxy. These rays constitute the preponderant source of the naturally created radioactivity that constantly bathes our planet. Some of these rays are so penetrating that they literally plow right through the entire earth and emerge out the other side to continue their travel through space. In addition to this natural effect, a transatlantic cable might also find itself near a drum of discarded nuclear waste, or an outcropping of radioactive uranium-bearing rock.

To learn what effects, if any, such radiation might have on the useful lifetime of glass fibers, I called a friend of mine, an expert on the subject. One input to the rough calculation we wanted to make was the amount of glass contained in a given length of optical fiber. She remembered "twenty-seven grams per meter" as the weight per unit length, but that didn't make sense to me. We'd both handled fibers very often and I couldn't imagine that a one-meter length of fiber would contain enough material to fill a shot glass.

A shot glass contains one liquid ounce, or about one-sixteenth of a pound of liquid. Two pounds (or pints) of water make up a quart, or 32 ounces, which is roughly equal to a liter. Since a liter weighs a kilogram—or 1,000 grams—dividing 1,000 (the grams) by 32 (the ounces) told me that the contents of a filled shot glass would weigh something like 30 grams. Since glass fiber is denser than water I figured it might take about half as much glass fiber to make the same weight of liquid—but still far more than one meter's worth. Once I made the connection, my friend quickly corrected her recollection to "twenty-seven grams of glass in a *kilometer* of fiber." The difference is 1,000 times.

My convoluted set of connections sounds really complicated, but that's more or less how I figured out that the original number had to be wrong. Since I'm so used to dealing with chains of numerical relationships, I completely overlooked a much easier way—one that only occurred to me as I was writing up this story. I needed simply to start (mentally) coiling up some fiber in a tight enough loop to fit into an envelope, and stop when I felt I had enough to need a second stamp. Since I had often handled optical fiber, it would have been easy for me to "see" that the coil would contain many dozens of turns—each about a foot long—so a single meter couldn't possibly weigh an ounce.

While the brute-force approach I took the first time around lacked the simple elegance of a comparison with first-class postage, it led me to the answer I needed. Both methods worked. The trick was to scan through familiar things with similar properties in order to build a path to the object in question. In order to get these, however, I had to make a number of rough estimates. Estimation gives problem solving the benefit of the imprecise knowledge our minds gather through everyday experience.

In most of my rough calculations, I treat the first digit as a 1 or a 3 and all the other digits as zeros. I like using three because it simplifies multiplication: 3 times 3 equals 10 in my system. Multiplying any given number by 10, 100, 1,000, or 1,000,000,000 can be accomplished by shifting the decimal point one, two, three, or nine places to the right, respectively. Similarly, division moves the decimal place to the left. In other words, it's simply a matter of adding or subtracting zeros.

When my son, David, was looking for his first job, one interviewer asked him, "How many barbers are there in the United States?" Not every engineering graduate would welcome that kind of question, but it was a good way of finding out whether or not a prospective co-worker could deal with the kinds of things not taught explicitly in engineering school.

David remembered that there were four barbershops on the main street of our town with a total of about ten barbers. Since our town has something over 10,000 people in it, that worked out to one barber per thousand, or about 200,000 barbers in a nation of 200 million people. The interviewer used a different method: one haircut per month for each of the 100 million people who get haircuts, and 400 haircuts per month per barber, which works out to 250,000.

When David presented me with the problem, I used money instead. I figured that each of the 100 million men who patronize barbershops spends $100 a year on haircuts, or $10 billion all together. I guessed that each barber must take in about $30,000 a year, in order to make a living and pay for a share of the shop, which gave me about 300,000 barbers.* The actual

*Culturally, I seem too far removed from this business to make accurate estimates. Men in New York, I'm told, often pay $25 per haircut. Apparently some barbers make quite a bit more than I imagined.

number is just under 100,000, according to the U.S. Department of Labor.

While some chains of connections miss the actual answer by a wider margin than others, there is only one *wrong* method for dealing with such problems: trying to memorize answers to all the questions one might be asked. My son's interviewer was looking for someone who could find ways of probing an unfamiliar idea, rather than someone who relied on memorized facts alone.

Science

Progress depends on new ideas from inquiring minds. But not every idea is worth pursuing. Ideas need testing and refining in order to establish their validity. The scientific method offers us a particularly useful tool kit for testing ideas. While most people aren't employed as full-time professional scientists, we all need the ability to demystify the world through the creation and testing of hypotheses. That ability lies at the heart of all scientific progress.

My daughter's dog fails to greet me as I enter the house. "Where's Harvest? She isn't out in the yard because I would have heard her barking. I'll bet Laurie took her for a walk . . . No, her lead is here on the hook. But the short leash is gone. Laurie must have taken Harvest in the car someplace. I think she mentioned needing to go to the veterinarian. Maybe she left a note." Test. Modify hypothesis. Test again.

The horn-shaped antenna Bob Wilson and I used for our detection of cosmic background radiation is located atop Crawford Hill in the suburban town of Holmdel, New Jersey. Geologically, the "hill" is just a pile of sand left behind by a glacier that visited our area about one million years ago, but it's high enough to offer a clear view of the surrounding countryside— including New York City's skyline some thirty miles to the northeast. While the horn had proven to be a far more precise instrument than its original design specifications had called for, it wasn't suited to the satellite communications and radio astronomy studies we wanted to carry out at much shorter wavelengths. As a result, we built a new antenna in the early 1970s to coincide with the launching of AT&T's first Comstar satellite.

Shortly after the new antenna was completed on the same hill, a Holmdel neighbor began having trouble with his television set. For no apparent reason, the picture on one of his channels would break up in the middle of a program—even though all the other channels were fine. It didn't happen very often, so he paid scant attention at first. But the effect persisted. Maybe it had something to do with the new antenna? He called Bell Labs, but our community relations representative assured him that none of the antennas on Crawford Hill had transmitters, they only monitored signals from space—signals that were far too weak to be picked up by a TV set.

Since none of his neighbors had similar trouble, he got his set checked, but without result. What was left? The next time it happened, he ran outside to see whether anything unusual was going on. Apart from normal traffic, and some activity around the antenna, nothing caught his eye. But he persisted. A few more trips outside renewed his suspicions about the antenna. There seemed to be something going on over there every time his set misbehaved.

His next call to Bell Labs was more insistent—he wanted to speak to the person "in charge of the antenna." That's how I found myself on the phone with him. At first, I tried to explain that our antenna couldn't be the source of an interfering radio signal, since we weren't sending out *any* signals. "But I've looked into every other possibility," he said. "And how come it never happens when all the lights at the antenna are off and there's no one around?" That's when I decided we'd better have a look ourselves.

When Mike Gans, one of our engineers, arrived at the house, he found the neighbor's TV set working perfectly. "Get the antenna to do something," our neighbor urged. Feeling a little foolish, Mike called up his friend in the control cab and asked to have the antenna moved. "Moving up in elevation," his friend reported. "Ten degrees, twenty, thirty . . ." At "sixty," the picture began to flutter. At "seventy," it washed out completely, and then recovered as the antenna passed "eighty." Dumbfounded, Mike had the antenna moved in the opposite direction. Sure enough, as the antenna moved through the earlier position, the picture broke up again.

Confronted with the evidence, we had little trouble locating the "interfering transmitter." In those days, northern New Jersey got its TV broadcasts from New York City's Empire State Building. (These transmitters have since moved to the higher World Trade Center.) While receivers with a clear line of sight to the transmitter get the best reception, most viewers can get acceptable reception from reflected signals that bounce around obstacles. Our neighbor got an extra bounced signal from the Empire State Building—via our antenna—which exactly canceled the one his set depended on.

To get this interference, the two signals had to be exactly the same size and their paths had to differ by exactly one-half of a wavelength—so that the "crest" of one wave would arrive at the same time as the "trough" from the other. The odds against such a coincidence were so high that we never even considered the possibility.

The remedy was quite simple. We moved our neighbor's TV antenna to the other end of his roof. That move sufficiently changed the relative path lengths of the two signals to negate the exact cancellation. Had the original installer picked that end of the roof in the first place, the problem would never have come up. All's well that ends well. Our neighbor got his reception back, and we got an unexpected example of the successful application of the scientific method—from a non-scientist.

In addition to solving puzzles, science also builds understanding by revealing the properties of the world and the relationships between them. Here again, the methods that scientists employ find widespread use in everyday life. From infancy onward, each person measures and classifies the properties of unfamiliar objects in order to integrate them into a larger worldview—from a ten-month-old learning to stack blocks, to Charles Darwin cataloging specimens aboard the *Beagle*.

My favorite examples of the scientific pursuit of knowledge come from antiquity. Almost 1,800 years before Ferdinand Magellan's ships sailed around the globe, an astronomer named Eratosthenes of Cyrene conceived and carried out the first accurate measurements of the earth's circumference some twenty-two centuries ago—but without leaving the African continent.

Simple observations will confirm that the world isn't flat. If you spot a ship at sea, and are sufficiently sharp-eyed, you will observe the ship "sinking" below the horizon as it moves away. At the same time, however, one can't use this effect to measure the curvature of the earth's surface reliably. Atmospheric effects blur and distort such distant images. Eratosthenes and his fellow astronomers could watch the steady motion of the sun, moon, and stars across the heavens break up into erratic fluctuations as these objects touched the horizon. The "sinking" ships might be some form of optical illusion, like a mirage in the desert.

If the earth were indeed round, proving it required a different kind of test. Eratosthenes needed some other way to establish a connecting bridge between the size of the "globe" beneath his feet and some measurable phenomena available within his own realm of personal experience. From experiments with shadows cast by sunlit objects at long distances from one another, he knew that the sun's rays moved in parallel lines. Thus, if the sun were overhead at one spot on the globe, sunlight would be tilted from the vertical at other locations.

Eratosthenes also knew from past experience that the noon sun was exactly overhead in the Egyptian city of Syene at each year's summer solstice.* How did he know? Because it shone to the bottom of a very deep well. Eratosthenes had also learned that, at the same moment at which the sun shone into the well at Syene, the shadow cast by an obelisk in Alexandria (some 500 miles to the north along the Nile) showed the sun to be some 7.5 degrees from the zenith.

Eratosthenes set about measuring the exact distance between the two cities. He gathered a group of people and trained them to take steps of a fixed size by having them march back and forth in a courtyard. He organized teams, gave each team a stretch of the 500 miles to pace off, and collected the total. In this way, he obtained a fairly accurate reading of the distance. (The word *mile* comes from the Latin word for *thousand.* A pace—or pair of steps—is about 5 feet long, and 1,000 paces gives us a mile.)

From geometry, Eratosthenes knew that moving 1/48th of the circumference of a circle tilted the local vertical by an angle of 7.5 degrees, the angle equal to the one cast by the obelisk. As a result, Eratosthenes concluded that the earth's circumference was 48 times the measured distance of 500 miles, or 24,000 miles (more precisely 24,662 miles). The modern value is 24,857 miles.

A few years earlier, Eratosthenes's colleague, Aristarchus of Samos, had used similar geometrical methods to measure the

*The day each year on which the sun reaches its northernmost point, usually June 20 or 21.

relation between the size of the earth and the distances to both the moon and the sun,* describing his work in a celebrated treatise entitled "On the Sizes and Distances of the Sun and Moon." We are fortunate that this literary treasure survived the Moslem conquest of the Roman Empire. Aristarchus's treatise was reintroduced into Europe at the end of the Dark Ages and helped pave the way for modern astronomy. (Copernicus referred to Aristarchus in his own work, for instance, but deleted the reference before publication.)

Geometrical measurements of the solar system improved as better instruments were introduced, but the principles remained the same for almost twenty centuries until supplanted by modern technology in the summer of 1962 when the first radar signals bounced off the sun's surface and gave us a more accurate value.

When I described Eratosthenes's and Aristarchus's measurements of the earth, sun, and moon to a colleague at lunch one day, I ended up covering the back of a menu with lines and circles. With my explanation finished, she said, "You know, geometry sits right behind the eyeballs!" I very much agree. Geometric ideas have a dominant visual component that still appeals to the human intellect—knowledge for the sheer joy of knowing. Unlike machines, the human mind can both create ideas and enjoy them as well.

Although primitive by modern standards, the technology of the time played an essential role in Eratosthenes's and Aristarchus's exploration of ideas. Throughout the ages, technology has helped shape the facts we humans think about. As our

*Aristarchus measured the size of the earth's shadow on the moon during an eclipse (with the earth between the sun and moon). He estimated the diameter of the moon to be one-third that of the earth, an error of only 8 percent. He found the sun to be twenty times more distant than the moon. While the actual value is almost four hundred, his value provided humankind's first inkling of the solar system's enormous (by terrestrial standards) size.

knowledge has increased, so have our tools and the ways we employ them. Today, technology is so complex and pervasive that it dominates much of the environment in which human beings live and work. For this reason, I feel we need a better understanding of how technology affects the ways in which we now create and explore ideas.

Chapter 8

Implications

*More perhaps than machinery, massive and complex business organi-
zations are the tangible manifestation of advanced technology.*
—JOHN KENNETH GALBRAITH, *The New Industrial State*

GALBRAITH WROTE those words in the pre-com-
puter era,* when America's Fortune 500 companies together employed less computing power than some of
today's single-user desktop machines. In those days,
"office electronics" meant an intercom between boss and
secretary, while "factory information systems" depended on clipboards and pencils. Computers had barely passed the laboratory curiosity stage and were rarely
seen outside payroll and accounting in even the most
venturesome organizations. This was, it must be noted,
a mere several decades ago.

Since then, computerization has expanded the role of
technology dramatically, particularly in the movement
and management of information. On a typical working
day in the late 1980s, for example, computers produce
over six hundred million sheets of printout in the United
States alone—some thirty times the document output
of the human workforce. Computers are everywhere.

*Work on the book began in the late 1950s, but was interrupted by
the author's government service during the administration of John
F. Kennedy. The book was published in 1967.

Many of today's offices contain more microprocessors than people. This picture of today's workplace brings me to the implications of technology and people at work.

How has the advent of electronic computing affected the validity of Galbraith's observation? Does the addition of information technology lessen or reinforce the organizational complexity that earlier technologies demanded? What role does individual expertise play? What about managers? Will their role be undercut by computerization? Finally, what about the technology itself? How can human organizations best apply knowledge in the creation of products and services? These are some of the questions I hope to answer here.

Complexity

In my experience, new companies generally start "simple." In a typical start-up, a small group of people organize to exploit a market niche—filling a need that others haven't recognized. The world doesn't look all that complicated at this stage. If the product meets its original goals, and customers like it, one must merely get the product out the door and to the customer. But I can assure you that things rarely remain that simple for long. Customers soon uncover shortcomings and demand more features, while competitors wake up to the new opportunity and begin to move in—often with more advanced technology. The start-up must scramble to handle these problems. Each new situation calls for expert help, help from people with specialized expertise. As these specialists are hired and integrated into the organization, the start-up takes on the complex attributes of modern corporate life.

On January 3, 1977, for example, Stephen Jobs founded Apple Computer with his partner Steve Wozniak, the only other employee. Ten years later, this two-man start-up had grown to a company with over five thousand employees and some $2 billion in annual sales, but both Jobs and Wozniak were missing from the corporate lineup. In the words of one of its engineers, Apple had become "just another big boring company."

Jobs hadn't planned it that way. Early on, the breezy informality of his managerial style created a team of highly motivated free-spirited people who worked to make the company and its products successful. As Apple grew, however, so did the "team." Specialization became necessary, and with it a need for more coordination. New teams split off from the original one in the early 1980s, and some tasks (such as the design of technically advanced computers) got done twice, while others (such as continued support for existing products) went neglected.

Apple produced two unsuccessful new products, the Apple III and the Lisa, one right after the other, and began losing ground to IBM's superior marketing. Sales, profits, and the price of Apple stock moved downward with no clear end in sight. Jobs realized he needed experienced managerial help and, after a careful search, recruited John Sculley from Pepsico in mid-1983. Sculley, as president of Apple, was to systematize the organization and to run the company's operations, leaving Jobs, who became chairman of the board, more time to concentrate on an elite team developing the company's hoped-for turnaround product, the Macintosh.

Sculley went to work centralizing organizational functions, instituting financial controls, and formalizing management structure—thereby ensuring proper attention to each of the factors involved in Apple's success. ("Hiring more MBAs" was the way an Apple employee described the process to a reporter.) Sculley emphasized marketing and set down rules to guide relations with dealers and independent software vendors. He put more work into upgrading the highly profitable Apple II, and "opened" the architecture of the Macintosh so that customers could connect it to other vendors' equipment. As a result of these new procedures and design standards, most Apple employees found themselves spending more time checking with others and less time having fun in their accustomed modes of operation.

Within a few months, Sculley's reforms began to slow the downhill slide, but reversing it took longer. The business didn't bottom out until 1985 and only then began moving upward. While still a distant second behind IBM, Apple's products began finding their way into business markets. Eventually, this thrust did well enough to earn a cover story in *Business Week* entitled "Apple's Comeback," but the going wasn't easy. All through 1984, Sculley continued to seek more help by creating new management slots and filling them with experienced executives.

The new management team's efforts strengthened sales and profits, but Jobs became increasingly disenchanted with the changes taking place. The final break came early the following year when Sculley named Jean-Louis Gassee—a successful veteran of Hewlett-Packard, Data General, and Apple's own overseas operations—as marketing manager of the Macintosh division, Jobs's personal pet. Jobs opposed the move, but at that point he owned only 11 percent of the stock and was outvoted. He left the company in September of 1985, described by *Fortune* magazine as a "champion of innovation" and a "foe of bureaucracy."

As the company celebrated its tenth birthday in January of 1987, John Sculley stood as its unchallenged leader. With Apple stock selling at three times its 1985 low, and sales and profits comfortably ahead of pre-slump days, the occasional grumbles of "old-time" employees provided the only sour note to the transition from entrepreneurial leadership to systematic management. Apple had grown up by growing the integrated organizational structure it needed to compete in a complex environment.

To me, the story of Apple epitomizes the conflict between individual autonomy and organizational success in a high-tech world. Understandably, Jobs wanted as much of both as possible. Once Apple began to grow, however, Jobs needed help to run the company. The issue was how much freedom this "help" would leave him. Jobs hired Sculley to handle "organizational" matters so that he could remain free to focus on creating new products—the thing he had enjoyed doing when the company was smaller and untroubled by competition.

Apparently, the part of the business Jobs wanted to keep for himself couldn't perform its function in isolation. When Apple had been the only game in town, the engineering team could design a machine to please themselves. But competition, and the need for larger sales volumes, raised the stakes and restricted design freedom. Under the old regime, Apple engineers might have been free to create the complete stand-alone system they thought personel computer users ought to have. But the new management needed products that fit the needs of increasingly fussy customers. That led Sculley to appoint an experienced marketing manager to the Macintosh design team, and the rest is history.

While Jobs left his company when it "grew up," other equally successful entrepreneurs have succeeded in making the transition. For example, Digital Equipment's founder, Kenneth H. Olsen, manages a company with over one hundred thousand employees and over $10 billion in sales. Despite the unflattering pictures of "bureaucracies" that most of us share, the fact remains that entrepreneurial owners spend their own money to build complex organizations.

In my own job at AT&T, I see a continuing need to increase links between people in various parts of the organization to keep pace with technological advances. As foreign and domestic competition drives us to seek constant improvement in our operations, designers can no longer afford to wait until a product emerges from a factory before we find out how well it works. Today, we do a far more integrated R&D job, designing and producing more complex products at lower costs and in less time than in the past. Today, our engineers and product managers must telescope product specification, functional design, design for manufacture, factory layout, and product testing into heavily overlapped schedules. Thus, much of the work goes on at the same time and puts a huge premium on coordination.

Like most of its competitors, AT&T needs fewer people to do a given job than it used to. Better coordination eliminates duplication and reduces the amount of rework needed to fix mistakes. But slimmer organizations aren't necessarily simpler. In my experience, an organization's complexity has less to do with absolute size than with the amount of dependence between its various functions, and the number of people who hold stakes in a given decision.

When I joined Bell Labs in the early 1960s, the product-realization process was a sequential series of information hand-offs, plans "mailed" from one group to the next. Today, everyone works from a single design plan that takes the manufacturing process into account from the start. This growing level of interdependence shows itself in the integrated way we now design and produce the things we sell. The extra care pays off by avoiding costly surprises later, but it means attending a great many meetings.

Today, most of the major manufacturing companies I'm familiar with have a mix of excellent, good, and not-so-good factories, but the standard keeps moving upward. Today's "good" factories would have been judged "excellent" just a few years ago. Today's best factories operate with a mechanical precision that in, say, 1985 I would not have believed possible (or even desirable). Most operations mesh so tightly with their neighbors that an entire production line can come to a halt if a single machine breaks down.

Until the Japanese taught the industrial world otherwise, factory designers kept their production lines loosely connected. Conventional wisdom called for "buffering" operations by providing storage space between each pair of adjacent stages in the process. That way, if a particular machine stopped for a while, its upstream and downstream neighbors could keep working by filling and emptying their respective buffers.

While this seemingly sensible arrangement kept everyone "busy," it also produced some unpleasant side effects. For example, when a particular machine went out of adjustment, it produced bad products until someone downstream discovered the problem. But that didn't happen until all the older material stored in intervening buffers had been exhausted—and replaced by an equivalent amount of newly made junk. The bigger the storage, the higher the cost of each malfunction. The desire to keep everyone else "busy" when one contributor got into difficulty created problems that we can no longer afford. Modern production lines must generally be able to handle a multiplicity of models and shift smoothly from one to another. Imagine what would happen if *most* of the machines in an assembly line shifted over to another product and someone with a supply of items from an earlier job shipped this now-unwanted material downstream.

Until recently, most of us accepted such glitches as the inevitable waste products of mass manufacture, the best one could achieve from the sum of the available (individually optimized) parts. Now, however, competition has forced electronics producers to stretch toward levels of quality and efficiency achievable only by near-perfect integration of effort across the entire enterprise.

When Japanese products first entered the American consumer electronics market, U.S. manufacturers were caught unprepared. Many blamed their troubles on "cheap overseas labor." In reality, however, most Japanese advantage flowed from better-engineered factories, which turned out higher quality products at lower cost. American manufacturing methods in this field proved no match for their competition, and by the mid-1980s every major U.S. producer of consumer electronics was importing virtually everything they sold from overseas. While some TV sets, radios, tape recorders, and stereos still bore familiar American names—RCA, Sylvania, GE—the small print on their labels invariably indicated foreign manufacture.

On the other hand, some American manufacturers have increased their share of both domestic and overseas markets. Black and Decker provides a notable example. In the early 1970s, Black and Decker's management faced the need to redesign its line of power tools to provide "double insulation"—an additional insulation barrier to protect users from shock if the main insulation failed. While some of the competitors had introduced double-insulated products at a 15 to 20 percent premium, Black and Decker's management resolved to use the new design to reduce the cost of their product line.

The integration of design engineering and manufacturing was the key element in achieving this goal. Manufacturing engineers from every part of the operation—machine and process development, cost, purchasing, and packaging—were brought in to work directly with the design engineering group. Organizationally, this meant combining manufacturing, product development, and manufacturing engineering under the newly created position of vice-president of operations. Through this combined effort, a small number of easily manufactured standardized components were stretched across the entire product line. For example, a single cylindrically shaped motor design supplied all needs. Only the length of the motor had to be varied to meet the different power requirements of individual tools. This permitted a single set of machines to produce all of Black and Decker's motors.

By standardizing on a much smaller number of components—such as fasteners, gears, and housings—volumes were increased to levels that justified internal production of specialized parts from lower cost materials. These new processes were also carefully integrated with design. In the case of gears, for example, bevel gears (cone-shaped gears that work in pairs set at right angles to each other and calling for precision machining) were eliminated in favor of spur gears, which could be mass produced from powdered metal.

The redesign process also called for a large capital investment in new machinery and the patience to wait for a long-term payoff—some seven years from the start to the break-even point in the original plan. When the payoff finally came, Black and Decker benefited handsomely. From 1967 through 1980 the cost of their products fell by well over 50 percent when measured in constant dollars. Had Black and Decker retained its 1960s' methods, the same products would have cost about five times as much to produce.

Black and Decker's integrated approach to the creation of its products led to a shakeout in the domestic power tool industry in which less prepared producers—Skil, Stanley, General Electric, Sunbeam, Thor, and Porter Cable, among others—did not survive. Black and Decker also gained a substantial share in overseas markets. As a result, you can find Black and Decker products in hardware stores around the world, labeled "Made in U.S.A."

Since winners and losers had access to the same technology, "winning through integration" was the name of the game and the better integrator came out on top.

At the yet higher level of integration demanded by the production of more sophisticated electronic equipment—such as computers and telecommunications gear—high-quality manufacturing is only part of the story. Producers of these items must be prepared to compete across a broad range of product requirements. An electric drill or TV set can be used just as it comes out of the box. A modern telecommunications switching system, on the other hand, must be customized to meet local traffic needs. It must be able to handle whatever equipment and signaling and networking arrangements the customers have been using. Training, maintenance, and a path for future expansion must be arranged in advance. Similarly, the large mainframe computers that store and control massive data bases call for equally complex care and feeding.

Since computing and telecommunications share much of the same technology, as well as portions of the same task (the movement and management of information), many people imagined that expertise in one field would transfer easily into the other. The news stories following the 1982 breakup of the Bell System provide notable examples of such expectations. Following the breakup, the *New York Times, BusinessWeek, Fortune,* and many other publications heralded the "Battle of the Giants"—the struggle of AT&T (the "telephone" company) and IBM (the "computer" company) to move into each other's business. I particularly remember the words of one writer who likened this "struggle" to a wrestling match between two "800-pound gorillas."

The actual event has proven far more mundane. While both companies have made attempts to invade each other's territory, nothing much has happened. It simply proved tougher to get into the other business than the pundits had expected. Despite the formidable resources available, neither one could duplicate enough of the other's organized expertise to make a dent in their core business.

Why? Basically, the answer lies in people. These "high-tech" businesses call for the utilization of specialized knowledge in highly sophisticated areas that go beyond the hardware itself— such as systems engineering, the management of large software projects, human factors, and operations support, as well as marketing, sales, and distribution. Galbraith's "massive and complex business organizations" continue to provide our best means for integrating diverse knowledge inputs and creating a synergistic whole. As I see it, the integration of individual expertise is still the name of the game, and the game still drives the major players to ever greater complexity.

Managers

At the same time, the imperatives that drive organizational behavior toward interdependence also create more work for the middle managers who must establish and maintain needed links between cooperating groups. In addition, as automation reduces the number of lower level workers needed to perform a given task, it generally demands more work from the managers that supervise them. As I see it, these growing needs for the services that middle managers provide are the key driving forces behind the dramatic changes taking place in the employee mix of information technology companies.

Over the past ten years, for example, a vigorous automation program has allowed the Traveler's Insurance Company to double the amount of business it handles without increasing its workforce. At the same time, the clerical component of that workforce shrunk from two-thirds of the total to one-third, while the professional/managerial component doubled in size.

Standardizing Traveler's Insurance's computer systems and networking them together have reduced the need for manual reformatting of information—such as typing names and addresses found on one form, or in one data base, into another. That cut the need for clerical help. At the same time, developing and maintaining these integrated systems called for highly trained people, such as computing professionals and systems analysts. Moreover, the increased business volume brought with it a proportionate need for more people to deal directly with customers and handle the nonroutine components of the workload.

As many of the cases I touched upon illustrate, machines can't replace people, they can only take over some of the work that people do. In the following hypothetical exercise, I'll try to demonstrate how automation shifts human workers toward jobs that call for more supervisory attention by taking over tasks that require relatively little managerial supervision.

Let us suppose that you direct the customer service office of a computer company. Your staff consists of forty clerks and four supervisors. These particular clerks spend half their time responding to pricing inquiries, which require nothing more than looking up items in a catalogue, while the other half of their time goes to answering relatively complex technical questions. The clerks handle most of these technical items by themselves, but normally have to send about 10 percent of them to their supervisors for resolution.

Here is the problem: one day your boss announces that the introduction of a new product will exactly double the customer service workload. At the same time, the company will install a computer-based inquiry response system that will automatically intercept and handle all catalogue-related questions. "Since the new system will relieve your people of one-half of the doubled workload," the boss continues, "your office will end up with exactly the same amount of work. So you ought to be able to handle the new arrangement without trouble." How do you respond?

Do you agree with the boss, or do you ask for extra help? If the latter, what kind of help do you want? In theory, freeing your clerks from the need to handle routine inquiries frees up half their time, so each one can do twice as much technical work. But human beings aren't as interchangeable as machines.

Some clerks might find a full day of technical problems too demanding, for example.

Do you ask for extra supervisors? In the old system, each clerk sent one-twentieth (one-tenth of one-half) of each day's work up the supervisory line, so the forty clerks gave the four supervisors a total of two days' worth of technical work every day. That left half of the supervisors' time available for supervisory tasks. With the doubling of technical problems, however, each group of ten clerks will send up another half day's worth of work, corresponding to a total of two extra days' worth of technical work for your supervisors.

Adding two extra supervisors and earmarking two-thirds of each of six people's time to answering technical questions, and one-third to supervision, would add up to the correct totals, but there's a catch. Six supervisors who each devote one-third of their time to supervisory activities can't get as much done as four half-time supervisors. They must attend the same number of meetings, for example—and adding extra people to those meetings will generally make them longer instead of shorter.

Instead of adding even more supervisors, I would add a couple of technical problem-solving specialists to the group, and give them the extra work. These specialists would provide the clerks with needed backup without giving you, the director, more line managers to deal with. It also provides a way of promoting good workers who have a different mix of skills than the ones that supervisors need. To cast the problems in terms of specific numbers, I had to make the problem artificially abstract, but the same principles operate in the real world. Through automation, therefore, I think it's safe to say that the infusion of computer technology into the routine aspects of work expands the role of middle managers.

In addition, our growing appetite for technology continues to raise the sophistication and specialization of our workforce. As this process continues, I see more managerial effort going into understanding the work of subordinates in order to direct their work. The more knowledge a job demands, the more help and expert supervision it requires. In my own job, Bell Labs research scientists provide a clear example of this need. Because much of their work is exploratory, subject to unpredictable circumstances and largely driven by self-motivation, one might think that the best management can do is to leave them alone. In fact, however, quite the opposite is true.

Most of the research scientists I know complain if their supervisors don't understand their work in good detail. As a result, much of each research manager's time goes into keeping up with technical developments within the organization, as well as with related activities elsewhere. The insight and guidance that these managers provide play a vital role in the success of the research process. I think it's fair to say that similar needs exist in most work situations that involve specialized expertise.

At Bell Labs, we naturally try to organize our work around group leaders whose expertise encompasses the primary field of the group's activity—or at least a closely related one—whenever we can. For example, about a dozen of our physicists work to tailor the properties of the semiconductors we use to make transistors, light-emitting diodes, and similar devices by slamming fast-moving atomic particles into their crystalline surfaces—a process called ion implantation. But that apparent homogeneity breaks apart on closer examination. Each researcher works within a unique portion of this subspecialty, and each requires individual attention from management.

For all of Bell Lab's involvement with technological change, the need for middle managers has remained remarkably stable. In fact, our organizational structure differs little from the one the Israelites adopted on their march from Egypt to the Promised Land—"leaders of thousands, leaders of hundreds, leaders of fifties and leaders of tens" (Exodus 17:21)—except we call them executive directors, directors, department heads, and supervisors. The structure resembles a pyramid, with most of the weight near the bottom, and progressively leaner toward the top.

While technology has generally increased the management workload, it also offers help for management by providing easier access to information—as well as personal aids to information processing. This latter area includes a variety of computer-based "decision-support" systems. These software packages handle information-processing chores, such as tracking financial performance, scheduling airlines, or laying out a network of telephone cables, that highly skilled people (both managers and non-managers) formerly accomplished manually.

My favorite example of an excellent decision-support system comes from a Sumo wrestling match I happened to watch during a recent trip to Japan. An elaborate ceremony preceded each match, after which the contestants tried to butt or throw their opponents out of a small ring. The first one to step outside the ring or touch the floor with any part of the body other than the soles of the feet lost the match—often by a fraction of a second.

While little in Sumo wrestling has changed for centuries, a key piece of technology has recently been added. Before making a close call, the referee—an elderly gentleman with lacquered headgear, long pointed mustache, and an elaborately embroidered silk robe—and his fellow judges now consult their "assistant," a microprocessor-controlled video recorder.

For me, the picture of those judges gravely checking through the frames one by one captured the essence of the proper interplay between the human and mechanical contributions to decision-making. In many ways, the Japanese appear more willing to incorporate new technology into their everyday lives, but as this case illustrates, they also maintain a good appreciation of

the appropriate division of labor. Refereeing is, in a sense,
a middle-management activity. Technology enhances the decision-making process by helping people do higher quality
work—making the human decision-maker's services more valuable.

Quality

As the pace of technological advance complicates our environment, it also leverages the impact of individual decisions. Small
mistakes, such as an automobile throttle linkage that sometimes
sticks when dirty, can lead to large consequences—serious acci-
dents and the recall of millions of automobiles. Consequently,
the productive power of technology forces us to shift ever more
of our attention from quantity to quality, from how much we
produce to how well we produce it—the whole, rather than the
parts.

In frontier days, pioneers often burned their houses down
when they got ready to move. They wanted their nails back.
The cost of a handful of component parts dominated the value
of the entire structure and the work that went into building it.

Things are a lot different today. With the exception of a few
items that we haven't learned how to mass produce—such as
gem-quality diamonds—individual parts owe most of their
value to a role in some larger system. We rarely notice parts
unless their absence or malfunction causes the system to fail,
like a rubber gasket on the *Challenger*'s rocket. Failure can
make a particular component very expensive indeed.

In a world where piece parts cost little, I see the critical
need—as well as the greatest added value—to rest with the
smooth function of an entire system, one in which all the parts
work together in a trustable fashion. True quality in a modern
system demands more than the narrowly defined quality of its
individual components. The quality of the *Challenger*'s gaskets

wasn't the issue. The gaskets behaved exactly as carefully made pieces of rubber of that shape and composition could be expected to. They maintained airtight seals as long as they were kept warm enough to remain soft and pliable. Unfortunately, an unforeseen combination of chilly morning air and the flow of ultra-cold gas from the rocket's cryogenic fuel tanks changed the ground rules, and disaster struck.

In the face of the *Challenger* disaster, NASA's program was placed under careful scrutiny. Our society has yet to demand such a painstaking approach to quality throughout a variety of complex systems in many other fields. In particular, I'm continually struck by how rarely information gets to all the places where it's needed. Sometimes the unmet need exists only a few feet from an unused potential source, as in the following experience.

As a parent, I've had my share of visits to hospital emergency rooms—screaming children, harassed clerks answering the same question over and over, guards pushing to get the double-parked cars moved on, and nurses weaving through the crowd looking for patients. What a mess!

The people in the waiting room want help as soon as possible. Since they have no idea when they will be taken, they keep asking the clerk. You can't blame them for not taking time out to look for a legal parking space when they might miss their turn. But is all that uncertainty really necessary?

Upon arrival, the admissions clerk records each person's name, address, and other data into a computer terminal. That entry goes on a priority list that the medical staff uses to handle the caseload. Why can't waiting patients see the list? All it would take is a monitor like those in airports, with people's names on it instead of flight numbers. Who wouldn't prefer information to uncertainty? With a decent estimate of the time available, people could step out long enough to park their cars—legally—or buy a crying child an ice cream cone from the snack bar. But the system keeps that information locked up on the other side of the counter. Only the clerk gets to see it.

Emergency medical service calls for a number of specialized contributions, such as first aid, transportation, access to the hospital, information acquisition, scheduling, nursing, medical attention, documentation, billing, and follow-up. When each activity focuses on providing "quality service" according to its own metrics, important efficiencies get overlooked. If the waiting room had a TV monitor, the clerk would have to answer fewer questions and the guard would have an easier time handling traffic. Most important, this addition would reduce the amount of trauma suffered by the patient, the *real* quality of the entire process. The trick is to get the vendors of the goods and services we depend on to focus on the issue of total quality.

When I began to think about such information shortfalls, I tried to find cosmic reasons for this phenomenon—like a lack of computer literacy or the cost of custom software. Upon reflection, however, I've settled on a much simpler answer. Shortcomings in the integration of information can crop up whenever attention to total quality lapses—even when the technology involved is as simple as a printed page of instructions.

Lewis Thomas's autobiography* provides a poignant example of such an isolation of information. One day in the late 1930s, while working as a junior resident in one of Boston's largest hospitals, Thomas learned of an interesting case from a colleague. A young musician had been admitted that morning with a history of chills and fever during the previous week. The patient's blood samples revealed malaria, a disease so unusual in Boston that many on the staff took specimens for further study.

*The Youngest Science: Notes of a Medicine Watcher, Viking, New York, 1983.

As the day wore on, a growing number of physicians and medical students came to the patient's bedside to observe this remarkable case for themselves. But all this interest didn't help the patient. The young man became increasingly drowsy as clumps of infected cells blocked more and more of his brain's blood vessels. He fell into a deep coma, and by evening he was dead.

Silently, the house physician left the group standing around the bed and soon returned with a copy of the medical textbook he had fetched from his room. Opening it to the chapter on malaria, he read the following passage to his assembled colleagues. "Any doctor who allows a case of malaria to die without quinine is guilty of malpractice."

The young musician was an admitted heroin addict. He had apparently shared a needle with an infected visitor from a tropical climate. While the treatment of malaria had long been part of medical training, none of the attending physicians did more than study the unexpected appearance of this disease on a wintery day in New England—until it was too late. Only then were the words in the book carried to the bedside.

For each of us, the consequences of information shortfalls can range from minor inconvenience to ultimate tragedy. Surely we can do better.

Human society can derive much-needed benefits from the quality environment that only an integrated sharing of information can provide. As I see it, we need an integrated approach to quality, one that defines and realizes the performance of each system in its entirety, rather than the small-scale behavior of its pieceparts. Technology that is measured by its total impact on the human beings it serves provides a worthy goal for the information age.

Chapter 9

The Human Element

At the top, life seeks expression through particular individuals.
—SIMONE DE BEAUVOIR

OUR WORLD grows more interdependent every day. Communications can take our words anywhere, and computers can bring the world's information to our fingertips. As technology spurs the growth of "massive and complex" organizations, we need to examine the effects of technology on individual human beings.

Caught between rising expectations and a diminishing store of natural resources, our society must look to a continuing flow of innovation to meet its needs. Where will the flow of innovation come from? I don't believe that we need to subordinate individual creativity to closely supervised groups of people working on small pieces of a fragmented task in order to make progress.

What kinds of human-machine architectures ought we to aim for in the workplace of the future? While centralization brings disparate activities together and makes them easier to manage, it curtails the individual flexibility that people want. Dispersion, on the other hand, provides for more individual control but leads to duplication in isolated islands of incompatible technology. I think we can direct technology to get us the best of both worlds.

This leads us to the role of individual leadership of the complex organizations that technology helps produce. Decisions made by computers and committees cannot displace the need for thinking and feeling human beings in positions of responsibility.

An examination of these issues provides the agenda for this final chapter of our story.

Innovation

Among major information age organizations, universities are the only significant group that doesn't fit the familiar hierarchical structure of corporate and governmental life. Instead, the members of most academic departments zealously guard their independence from the "administration" to which they nominally report.

Shortly after Dwight Eisenhower left the presidency of Columbia University to become president of the United States, his former colleagues received a telegram from the White House inviting them to send a "high-level" delegation to a White House conference. "Nothing lower than a dean," the telegram concluded. Columbia responded with seven words: "There *is* nothing lower than a dean!"—clearly separating themselves from the respect for hierarchy that pervades almost every other facet of organizational life in our society.

In contrast to their corporate counterparts, the research-oriented faculties of the world's major universities profess little need for organizational coordination. This difference reflects a fundamental split between their functions. Corporations strive to integrate information in the service of a common task—commerce. Universities, on the other hand, create knowledge in order to disseminate it. At present, universities organize to prospect for gold rather than to build a supertanker.

Not everyone agrees that this individually focused approach to innovation still makes sense. In recent years, for example, most advanced nations have launched centrally organized research initiatives, such as Japan's Fifth Generation Computer programs, Europe's Esprit, and the United States' Microelectronic and Computer Technology Corporation. Has high technology made successful individual researchers an endangered species? Personally, I don't think we have to look to big science for most of our future breakthroughs. An examination of three of science's most spectacular recent advances tells a story of science on an individual scale.

The 1986 discovery of superconducting ceramics set the scientific world on its ear. Half a century of superconductivity research had produced a handful of metals whose electrical resistance vanished when cooled near absolute zero. This early work was interesting from a scientific point of view, but led to few applications outside the laboratory. Instead, K. Alex Muller and J. George Bednorz had discovered a substance that completely lost its resistance at far higher temperatures. While earlier superconductors required cooling in liquid helium—science's most exotic and expensive refrigerant—some of the new ceramics* became superconducting when immersed in liquid nitrogen, a widely available coolant that costs less than beer. Each new result fueled growing speculation about economic benefits from the lossless transmission of electric power, high-speed trains levitated by superconducting magnets, ultrafast computers made from this new material, and a host of other applications.

*Ordinary ceramics don't conduct electricity at all. That's why power companies use ceramic insulators between high-voltage power lines and the metal towers that support them.

Within months of the initial announcement, well over a thousand researchers around the world had abandoned what they were doing and raced to explore this amazing new discovery. The speed of its impact was reflected in the unprecedented recognition it received. As a result of a hurried nomination that barely made the February 1 deadline, Muller and Bednorz were awarded the 1987 Nobel Prize in physics. Nobel Prizes normally lag discoveries by several years and often by several decades—Einstein himself had to wait sixteen years—but this issue was settled almost the moment the original discovery was confirmed.

The size of the effort needed to produce this scientific advance stands in sharp contrast to its huge impact. In reporting the Nobel Prize announcement, the *New York Times*'s Walter Sullivan reported Muller and Bednorz as "working with inexpensive equipment in a small room at the IBM Zurich Research Laboratory . . . a standard cooling flask, some volt meters, a personal computer." The ceramics they used were formed from a recipe created by two French chemists, Claude Michael and Bernard Raveau. This recipe's ingredients—and those of the others that followed—come straight out of a chemical catalogue, and they can be cooked together using a few hundred dollars' worth of standard chemical apparatus plus a household microwave oven.

The 1987 Nobel Prize marked the second time in two years that IBM's Zurich research had won this award. Gerd Binnig and Heinrich Rohrer's invention of a new kind of microscope earned them a share of the 1986 prize, together with Ernst Ruska of Berlin, who was cited for an earlier discovery. In 1931, as a student at Berlin's Technical University, Ruska found that a magnetic coil could focus a beam of electrons in much the same way that a glass lens focuses light. By 1933 Ruska managed to improve his apparatus—a beam of electrons moving through a pair of magnetic lenses and a detector—to produce the first electron microscope capable of revealing details smaller than those conventional optical instruments can resolve.

Some fifty years later, Binnig and Rohrer took a totally different approach to microscopy. They mapped the surfaces of crystals by moving an ultrasharp needle back and forth across them. An ingenious electronic control keeps the needle's tip a fixed distance—generally less than the width of a single atom—from the crystal's surface. The tip's up-and-down motion at this very closeness traces the hills and valleys of the surface with a fine enough resolution to locate even a single atom sticking out beyond its neighbors. In many applications—such as mapping the contours of a crystalline surface—the so-called scanning tunneling microscope reveals surface features more clearly than even the most advanced electron microscopes.

With fifty years of work behind them—and million-dollar price tags in some cases—today's electron microscopes are truly high-tech instruments. Building a competitive, high-resolution electron microscope requires a massive investment that few equipment manufacturers can afford. By contrast, Binnig and Rohrer's work permits scientists to leapfrog the technology of atom-by-atom microscopy with no more resources than moderately well-supported research scientists routinely expect: a lab, access to a machine shop, a technician or two, and permission to buy some equipment.

Constructing such an instrument at Bell Labs, for example, caused hardly a ripple. The only unusual item was a request to move the apparatus to an isolated storage shed—normally used to store lawn mowers—in order to escape the small but unavoidable floor vibrations found in our main building.

The modest scale of the 1986 and 1987 Nobel Prize–winning work is also evident in the work cited in 1985. In that year, Klaus von Klitzing won the prize for his 1980 discovery of the quantum hall effect. At low temperatures, electrons flowing through thin layers of semiconducting crystals in the presence of strong magnetic fields exhibit unexpected jumps in their current-carrying behavior.* The precise regularity of these jumps offers both a fruitful new way to explore the electronic behavior of semiconductors and a way of measuring the fundamental constants of nature with unprecedented accuracy.

Of the three areas of investigation described above, von Klitzing's used the most elaborate piece of equipment, a powerful magnet. Unable to achieve the magnetic field he needed with the magnets available at Würzburg, his home university, von Klitzing performed his measurements at a shared facility in Grenoble, France, which offered scientific visitors use of a large magnet. Subsequent to the initial discovery, however, other researchers found that better crystalline samples permitted study of this effect without special facilities. Here, too, the breakthrough fit within the single-lab environment.

A common thread of individuality links these achievements. Small is beautiful. None required massive organizational support or direction. At the same time, however, each of the scientists involved was closely coupled to the work of colleagues through the professional links that permit scientists to exchange ideas. Access to ideas makes all the difference. Whenever I've been involved with a university tenure committee, the single question of substance asked has invariably been: "Do you think Dr. —— will be able to find first-rate problems ten years from now?" Nobody has ever asked whether or not Dr. —— would

---

*Charged particles tend to follow circular paths in the presence of magnetic fields. The slower the particle and the stronger the magnetic field, the smaller these circles become. With *very* small circles, von Klitzing found, nature prefers certain sizes to others. When he tuned the field of his magnet to produce some of these preferred sizes, the current jumped.

be able to *solve* problems. The ability to solve problems comes with the standard academic tool kit; it's the selection of the problem to be worked on that separates interesting science from irrelevant mediocrity.

As a result, I have found that managing successful research depends far more on ensuring the flow of ideas between individuals than on rigid direction from the top. Even though Bell Labs employs over one thousand researchers, it's rare to find as many as half a dozen people working on the same project. Once a research idea proves ready for *development* (i.e., design for manufacture), a much larger number of people become involved and careful coordination becomes essential—but rarely before.

As we've seen earlier, sophisticated technology leads to organizational complexity. But technology builds on knowledge, and knowledge begins with the ideas that individual human beings create. Thus the technology to meet the future needs of our complex society will very likely owe its foundation to people who do their best work in groups of ones and twos.

Networking

As work becomes increasingly information intensive, I see organizational success depending more and more on giving each individual contributor needed information at the right place, at the right time, and in the right form. The degree to which this requirement can be met depends crucially on the information architecture used, the organization's "nerve system."

Ever since the advent of large computers, or "mainframes," many organizations have chosen to fulfill their information needs by keeping all data at a central location. In such systems, the map of information flow resembles a wagon wheel with its rim removed. The terminal at the tip of each spoke sends and receives data from the mainframe computer at the hub. The mainframe handles all data processing, storage, and retrieval, as well as whatever communications between terminals is permitted by the system. No piece of datum is supposed to get out of reach because there is no way of creating, modifying, or storing records anywhere but at the hub.

Central data managers usually strive for efficiency by keeping formats uniform, by offering users only a limited range of options to meet their special situations. Modifying the system to handle exceptions is generally expensive; some pieces of information can't be accommodated. (Imagine a worldwide insurance company starting to do business in Saudi Arabia, for example. Should the central system expand the "spouse" category to permit multiple entries?) The trouble with centralization is that information that doesn't fit the format doesn't get stored by the system. If no better alternative is available, it's often relegated to a pile of papers in a corner of someone's office.

Such a drive toward uniformity usually saddles everyone in the organization with a single all-encompassing data communications network whose characteristics are defined by the central management. As a result, many local common-interest communities find themselves forced to find "private" means of meeting their specialized needs, and proceed to set up networks with their own data formats and connection schemes.

Consider the purchasing agent whose desk has three terminals crowded on it: a vendor-supplied order-entry machine, a second one that runs the specialized software the purchasing department uses to do its job, and the dumb terminal connected back to central files. This last terminal is needed for information such as which employee is authorized to sign which kind of purchase order. The purchasing agent also needs access to other networks, but the crowded desktop doesn't have room for more terminals. The result is a proliferating number of isolated networks, few of which offer information access to other people in the organization.

Even if the central authority manages to prevent the creation of "private" networks inside the organization, the problem of information isolation isn't solved. It is just moved to the boundary of the organization. If the terminals at the periphery can't "talk" to other networks, information must move across the organizational boundary in paper form. What's printed out by the system on one side of the boundary must be typed into the system on the other side by someone sitting at a keyboard. This is slow and expensive. It also makes interrogation across the boundary very hard.

Under these circumstances, it's hard to imagine that even the best-entrenched central processing culture can do much more than fight a rear-guard action against the proliferation of quasi-autonomous data centers. This inevitable proliferation calls for a new information-networking strategy, one that recognizes local options but also avoids the duplication and lack of coordination that come with isolation.

Overcoming isolation by attempting to connect everybody to everyone else is both prohibitively expensive and unnecessary. A better solution is to facilitate the creation of local networks that are themselves networked together. In this scheme, individual information needs can be met by participation within common-interest groups tied together by their own formal or informal networks. This essentially amounts to ratifying what people prefer to do anyway. From the users' point of view it takes membership in several such user-defined networks to satisfy the multiple needs of any particular individual.

Personally, I make use of at least four: local traffic with my office staff; private communication within Bell Labs' management; radio-astronomy data swapping and processing; and electronic mail. In addition, our electronic mail system connects me with other networks that permit me to exchange messages with friends and colleagues around the world.

Fortunately, the different kinds of terminal interfaces that connect to these various networks can often be accommodated electronically. These days, a typical computer scientist's desk holds a single "smart" multitasking terminal that can play an appropriately different role in each of a number of networks at the same time—with an interface to each network through its own window on the terminal's display. Since the number of networks that can be accommodated is no longer limited by desk space, some of the common-interest groups that had to be handled manually in the past can also get the benefit of electronic assistance. Individual users can thus formalize their ad hoc "networks" as well as define new ones to suit their needs.

Networking management is a tougher job in this user-defined environment than it was under the centralized scheme. (It's usually easier to be a dictator, at least until the next revolution comes around.) It isn't easy, but neither is dealing with organizations whose "centralized" models of information architecture limit their view of reality to the data in their files.

Let's assume that you have been hospitalized. You fill out the top half of a claim form and give it to the admitting office. During your stay, the people treating you jot information on various forms, which the hospital's data-entry clerks type into the hospital's computer. Later, some other clerk sits down with your insurance form and copies in data from the hospital's computer. When the insurance company gets that form, one of their clerks types the data into the insurance company's system. If any piece of datum is omitted or miscopied, letters must go back and forth, envelopes opened, files called for, changes made, and records of these extra transactions sent to central files.

In the meantime everybody is kept waiting. You may have

been obliged to pay on the spot and then wait for a refund. Sometimes the long-awaited reply arrives with only part of the money, or no money at all. A mistake has been made. But where? Calling the insurance company gets you a clerk who can only look into the insurance company's computer to see what some other clerk typed there. This causes you to ask to speak to a "supervisor," someone who can dig out the originally submitted forms—not easy, and never without a wait. But you want an answer. You call the hospital, only to reach yet another clerk who can only tell you what someone else copied into the hospital's computer from the staff's scribbled records. Frustrated, you go back to the insurance company and complain about the runaround.

No wonder your insurance company needs an extra layer of people to deal with complaints. But even these complaint handlers are forced to work with thirdhand information—a retyped copy of a retyped copy. In their "centralized" environment, they can't access the hospital's records directly, let alone the doctors'. Everyone is constantly on the phone "straightening things out." Work jumps around from one job to another—depending on who calls back—and each new person you talk to requires reacquaintance with the specifics of your particular case.

Technology can ease these modern complexities. How much simpler to give your insurance company's people access to needed information from your doctor's records. This calls for a networked architecture in which the insurance company's front office has enough processing capability to provide the electronic interfaces needed to access the hospital's files. Each person in the insurance company's front office would have a "smart" terminal capable of creating several simultaneous virtual terminals out of software. One of these would be networked to the hospital's computer and behave just as if it were a terminal used by one of the hospital's clerks who usually fills out the forms, while a second virtual terminal would be connected to the insurance company's central computer—providing information on your policy's coverage, for example. In each case, your privacy would be assured by a verification system that certified your granting permission to have your records examined in this way.

The human operator would monitor the process and intervene as necessary—obtaining additional information and making corrections on the spot, instead of starting all over again after a round-trip through the mails. As a result, the multi-network architecture would allow a single member of the staff to handle your transactions with less hassle for everybody.

My wife developed a "tennis elbow" recently, even though she doesn't play tennis. Before proceeding with the treatment that her doctor had recommended, she decided to seek a second opinion from the osteopath that I use. She first had to take time off from work to pick up her X rays. Then with the X rays and insurance forms from both our companies, she went off for her examination. Afterward, the people in the osteopath's office made out a bill for us and filled out their parts of the two sets of insurance forms.

My insurance company bases what they will pay us on the difference between what Anne's insurance company pays and the total amount. As a result, they often write us to ask for a statement from the other company. Furthermore, some clerks save time by typing in only the first initial when processing a claim. Since Anne and I share the same first initial—as well as most medical and dental practitioners—the wrong person is often processed as the patient.

Sometimes one or the other company overlooks the fact that we have double coverage (even though we note that fact in the appropriate box on their forms). We sometimes pay the doctor and one or the other company does also, bringing extra bills, refunds, and acknowledgments into play. In a typical encounter, I've counted as many as two dozen different sheets of paper—half of them filed someplace—and as many as eight separate transactions over the course of several weeks.

In a networked future, the osteopath's assistant would get an access authorization code from my (paperless) wife at the time of her visit. Making the necessary calls on the spot, the assistant would dial up high-resolution images of the other doctor's X rays and any other relevant medical records, along with the status of our insurance coverage. Later, on the way out, the receptionist would do a bit of electronic banking with the two insurance companies, punching in the treatment codes and amount of fee. In unusual cases, a human clerk at the insurance company might come on-line to clarify the situation. If the insurance agent asked for high-resolution images of additional X rays or the osteopath's handwritten notes to document the claim, the osteopath's electronic image-storing system, or facsimile machine, would supply them in a few seconds. At the end of the transaction, the insurance company would transfer the appropriate amount to the osteopath's account—allowing the receptionist to immediately repeat the process with the second carrier for their part of the balance. With the two claims processed, the receptionist could give my wife an on-the-spot bill for the balance, a single sheet of paper after less than ten minutes of work.

Note that these networking examples *humanize* the patient's view of the process. Instead of dealing with a multitude of people, each with a fragmented view of reality—reality as viewed at the screen of an isolated computer—the patient sees a single human being, one with access to all the relevant information, and the power to act on it. I'm sure that all of us would welcome

that humanization of information architecture in all our transactions.

Leadership

Networked information access can realign fragmented tasks
into more meaningful work. But even the most sophisticated
technology is useless without human judgment and vision to
direct an enterprise toward its goals.

Technology offers vast new opportunities, but only if people
have the imagination to grasp them. Agriculture can surely
produce more protein as better technology is brought into play.
According to Joshua Lederberg, the president of Rockefeller
University, medical care stands on the brink of a revolution—
thanks to computer-aided probes of the processes that govern
life. Unmet needs abound in housing, education, and energy
use, which have barely begun to benefit from the impact of new
technology.

Certainly, today's leaders face formidable problems. But real
leaders find ways of overcoming obstacles that seem insur-
mountable to others.

Consider what Mao Tse-tung faced in creating the People's
Republic of China from hundreds of millions of inhabitants
scattered in hundreds of thousands of isolated peasant villages.
His enemies controlled the population centers as well as all
modern means of transportation and communication. Mao
managed nonetheless to create and maintain a loyal and disci-
plined organization whose members willingly endured decades
of unprecedented hardship in pursuit of a goal.

We needn't admire Mao as a person, or approve of his goal,
in order to recognize his ability as a leader. Because of the vast
differences in scale and in culture that separate the unique
events in China's history from our own experience, Mao's life

story serves more as benchmark than role model. Nevertheless, whenever some pundit proclaims our present problems to be "too big" to be managed, my thoughts go back to the handful of determined people huddled in the caves of Hunan Province who managed to redirect the lives of a billion human beings.

While successful business leaders lead much less spectacular—and less memorable—lives, their stories make for much more instructive role models. My favorite is a career corporate executive who became comfortably affluent rather than wealthy, enjoyed taking vacations, and never made much of a name for himself. His name was Theodore Vail, and he created the modern Bell System.

When J. P. Morgan brought him in to rescue an ailing utility in 1906, Vail faced a pack of problems. AT&T operated the majority of the United States' six million telephones, but only by a slim margin—a margin won by bruising competition and a costly acquisition program that left the company financially exhausted. Disillusioned investors had dropped AT&T stock to less than half its earlier peak. Customers angered by bad service, employees demoralized by poor supervision and unsavory working conditions, public anger at heavy-handed business practices, and a hodgepodge of barely usable technology all testified to the abysmal level of planning and leadership displayed by a management mired in internal power struggles.

With little prospect of borrowing more money, Vail issued new stock and offered it at a below-market price to existing stockholders (who were given the "right" to buy one new share for every six they already owned). Stockholders jumped at Vail's "bargain," giving AT&T some $20 million in much-needed cash and a restored level of credibility in the financial community.

At the same time, Vail quickly demonstrated his ability to spot and promote talented people, replacing the senior management's aristocratic pretensions with a new meritocracy. He created a consolidated R&D laboratory and placed it under his new chief engineer, John J. Carty, a self-educated former telephone operator, whose personal inventions had done much to improve the still imperfect art of telephony.

The new research lab quickly began to probe for ways to exploit Lee DeForest's recently invented "Audion" (precursor of the electronic vacuum tube) to amplify telephone signals and hence extend their range. While the Audion was far from a practical device, by 1909 Vail found the results encouraging enough to promise transcontinental service to all telephones in time for the 1914 opening of the Panama-Pacific Exposition. Fortunately, AT&T engineers came through in time, even though the practical vacuum tube needed to make this achievement possible was not invented until 1912, almost three years after Vail had announced the commitment and more than halfway to the 1914 deadline.

Vail's day-to-day moves flowed from the pursuit of larger goals, goals that he took pains to make clear to those around him, reversing the secretive managerial style that typified the business life of his contemporaries. He began his first annual personal report to stockholders with an essay on the relations between the corporation and the public, putting forth the heretical notion that maximum private profit was not necessarily the primary objective. Instead, profit was only one element in a more complex equation, involving long-term financial health, innovation, and service improvement. In this new concept the corporation aimed toward a proper balance of these factors. Vail saw regulation as a better alternative to competition, a "universality" of service that could not be furnished by dissociated companies.

During the early years of Vail's tenure, AT&T and its financial backers pursued an aggressive acquisition policy—merging independent telephone companies into the Bell System wherever possible. With better service and more openness in its dealings ("If we don't tell the truth about ourselves, someone else will"), AT&T's growth met with little resistance from the general public. By 1911, Vail had consolidated his collection of local companies and recent acquisitions into a set of regional companies covering most of the United States. Vail's goal—a "universality" of service—seemed within reach. With the election of a Democratic administration in 1912, however, the political climate began to change and the possibility of anti-trust action loomed as an obstacle to the "universality" of service that Vail sought.

Push on to finish the job? Rather than risk the breakup of what he had already built, Vail compromised. He halted the purchase of independent companies and agreed to provide them with arrangements for long-distance service, thereby including them in a shared nationwide network. The compromise worked. While other nations nationalized telephone service, AT&T remained out of the government's hands (except for a brief period in the waning days of World War I).

Vail retired in 1919 at the age of seventy-four. In little more than a decade, he had transformed a shaky company in a chaotic industry into the bluest of blue chips with a reputation for service of the highest quality. The Bell System he created survived intact for more than sixty years, and continues to thrive today even after its government-mandated split-up. Despite the passage of time, the advance of technology, and the consequences of political misjudgment, Vail's vision of "universal" service still guides the actions of the almost one million people who perform the complex array of tasks needed to make cross-continental conversations a routine convenience of everyday life.

Clearly, Theodore Vail didn't singlehandedly invent the idea of a service-oriented business ethic, nor did he discover the need for balance between private and public interest all by himself. But he proved able to embody his ideas in a powerful and highly visible example that others could adapt to their own needs. In that way, Vail's ideas moved beyond the boundaries of his company and helped shape the course of twentieth-century business life in the United States.

But leaders suffer setbacks, too. The company that Vail led to unparalleled success had actually fired him earlier. Vail's association with the Bell System began in 1878 when he left a promising (and better paying) career with the U.S. Post Office to become the first general manager of the Bell Telephone Company, just two years after Alexander Graham Bell's patent had been issued. With barely ten thousand phones in service, the Bell Company and its licensees were locked in a struggle with Western Union's larger (over fifty thousand phones) and better-financed effort. Furthermore, thanks to Thomas Edison's invention of the carbon microphone, Western Union's phones sounded better than Bell's.

Vail set to work with a vengeance, and soon the Bell System moved ahead of its bigger rival. Nevertheless, Vail found himself in trouble with the company's financial backers, who were less interested in his nationwide service ideas than in quick profits. Vail left the company, and spent a quarter of a century in other endeavors before J. P. Morgan's takeover brought him back.

Vail's focus on the future echoes a major present concern. It was only during his second term in office that Vail managed to convince AT&T's financial backers to subordinate short-term gain to more farsighted interests. That issue burns today. Eager investment analysts pick apart corporate balance sheets in order to feed the numbers to waiting financial markets. Lower earnings can mean a sharply lower stock price, unhappy stock-holders, and difficulties in borrowing money—and seed money invested for future payoff cuts into those earnings.

This dilemma provides a classic example of the perennial conflict between human judgment and the alleged objectivity of hard numbers. But earnings numbers take on meaning in the context of an organization's prospects. Investors generally pay attention to quarterly earnings as a means of charting a company's future. The president of a financially pressed corporation summarized his argument for a long-term view with one of my favorite one-liners: "They that live by the quarterly report shall die by the quarterly report." Decisions that consistently short-change the future in favor of more tangible gains will ultimately prove fatal.

The struggle between Vail's vision and the down-to-earth concerns of hard-headed profit seekers went far beyond money. It marked the beginning of a revolution in value—a revolution brought about by technological change. Vail's contemporaries measured value in terms of their prior experiences. Numbers of customers, dollars of revenue, and numbers of telephones fit well into a familiar value system based entirely on tangible things—acres of land, head of cattle, or tons of iron. Vail, on the other hand, recognized the tremendous value of an intangible item—universal connectivity, a company's ability to connect every customer to anyone that such a customer might want to contact. The advent of the telephone created a potential source of value different from that associated with the production of material goods.

As we move toward an increasingly information-based economy, value tilts toward intangible assets. Intellectual property—patents, designs, business plans, trade secrets, and computer software—is worth real money and must be protected accordingly. For many companies, however, their most valuable asset can't be kept under lock and key—the knowledge possessed by the individual members of their workforces. The need to husband this asset has spurred a drive to improve management mechanisms—productivity metrics, performance evaluation, compensation packages, career pathing, profit sharing, suggestion boxes, motivational seminars, and the rest. But unless the leaders of the enterprise act as real leaders, that impressive array of management tools can't bring forth the kind of self-motivation that productive knowledge work demands.

In the middle of a recent discussion of the future direction of Dell Labs' research, a friend remarked, "The reason we are all here is to make money for the stockholders." That observation took me aback for a minute. Is that really why my colleagues are so deeply involved in what they are doing? Clearly, the building we occupy, the equipment we use, and the salaries we earn all come to us from somebody who expects to make money—otherwise none of it would exist. But on further reflection I realized that "the reason we are all here" doesn't tell the whole story. That top-down view misses an important ingredient. We must never stray from the perspective of the individual, "the reason each of us *chooses to be here.*" For most of us, that turns out to be the opportunity to work with others, doing something we think is important.

There are walks of life that seem to offer little more than the gratification of avarice, like "playing" the stock market. But for many others, meaningful work strives to make a meaningful difference in some aspect of the world we live in. Making a meaningful difference on a global scale will take the best use of the minds and machines that our society can muster. Machines need direction from human minds, and human minds need inspiration from human leaders.

Index